Lecture Notes in Mobility

Series editor

Gereon Meyer, Berlin, Germany

More information about this series at http://www.springer.com/series/11573

Emma Briec · Beate Müller
Editors

Electric Vehicle Batteries: Moving from Research towards Innovation

Reports of the PPP European Green Vehicles Initiative

Springer

Editors
Emma Briec
DEA, Innovation Strategy and Planning
R&AE
Renault
Guyancourt
France

Beate Müller
Future Technologies and Europe
VDI/VDE Innovation + Technik GmbH
Berlin
Germany

ISSN 2196-5544 ISSN 2196-5552 (electronic)
Lecture Notes in Mobility
ISBN 978-3-319-12705-7 ISBN 978-3-319-12706-4 (eBook)
DOI 10.1007/978-3-319-12706-4

Library of Congress Control Number: 2014956707

Springer Cham Heidelberg New York Dordrecht London

Printed on acid-free paper

Springer International Publishing AG Switzerland is part of Springer Science+Business Media (www.springer.com)

Foreword

Battery research is at the heart of one of the most important transitions our world will have to face in the future. Transport and energy have always been strongly linked, but the emergence of electrification in road transport means that electrochemical storage technologies will play a stronger role in our cars. With the emergence of plug in hybrids and extended range electric vehicles batteries might not necessarily completely replace conventional fuels, but will still play a paramount role in this shift, and therefore Europe needs to recover a major role in this industrial domain.

European researchers have played an important role in the early development of lithium-based batteries, which are currently dominating the world market and will enable the current generation of electrified vehicles to provide more appealing range and performance to customers than their predecessors. These vehicles, however, in most cases are powered by batteries designed and built outside Europe. While at current sales levels this is not yet a major issue, European researchers and industries should use the time it will take to ramp up sales of electrified vehicles to bridge this gap, aiming to recover production to Europe by developing a new generation of high performance cells that rival performance with Asian and American products. This is where research funding plays an essential role, and why the European Green Cars Initiative (EGCI) dedicated 25 projects, for a total of more than 85 M€ to electrochemistry and battery management, as well as their integration.

A similar effort is dedicated to this sector in the current Horizon 2020 Research Programme, within the European Green Vehicle Initiative that follows the EGCI. The revised structure of this public–private partnership widens the coverage to new types of vehicles (from two wheelers to buses and trucks) and alternative energies. The EGVI package is intended to provide all stakeholders in the automotive sector an incentive to pursue decarbonisation and air quality improvement while at the same time developing a new path to world level competitiveness. We expect that electric batteries development and manufacturing will be a significant part of this future European success story.

Manuela Soares
Director for Transport
DG RTD, European Commission

Preface

An important instrument for supporting research on electrification of cars has been the European Green Cars Initiative Public Private Partnership (EGCI PPP) which was set up within the Seventh Framework Programme in order to fund research and demonstration projects on electrification, logistics and heavy duty transport. In Horizon 2020, the EGCI PPP is now succeeded by the European Green Vehicle Initiative Public Private Partnership (EGVI PPP) that focuses on energy efficiency and alternative powertrains. The initialization of a PPP gave the opportunity to build a close dialogue between the stakeholders of the industry, research institutes and the European Commission. This is among others reflected in the regular expert workshops that were a joint activity of the industry platforms European Technology Platform on Smart Systems Integration (EPoSS) and European Road Transport Research Advisory Council (ERTRAC) and the European Commission and prepared by the Coordination Actions "Implementation for Road Transport Electrification" (CAPIRE) and "Smart Electric Vehicles Value Chains" (Smart EV-VC).

This proceedings volume is a report on the scientific talks that were given on one of these workshops on the topic of EV Batteries: Moving from Research towards Innovation which took place on 10 April 2013. The aim of the workshop was to provide recommendations on R&D&I support activities in the framework of Horizon 2020 based on: a review of the results of collaborative research projects on batteries funded under the European Green Cars Initiative, a review of relevant attempts in implementation of prototype manufacturing and mass production in Europe and a discussion on current EU activities and policies for bridging the gap between research and innovation in the domain of batteries for EVs, including European activities and policies to foster innovation. Invited experts included the coordinators of European collaborative research projects on batteries, leaders of major pilot activities for battery manufacturing, as well as representatives of European companies active in battery technology, automotive manufacturers and—suppliers and research institutions. Representatives of relevant Directorates General of the European Commission also participated.

Currently, there are 25 projects funded within the European Green Cars Initiative PPP dealing with electric vehicle battery materials, technologies, processes and manufacturing. The scientific talks in the workshop focused on innovative battery materials, advanced manufacturing processes and smart battery management systems.

The purpose of this proceedings volume is to disseminate the results of the European Green Vehicles Initiative PPP to a broader stakeholder community and to further strengthen the dialogue among the stakeholders and with policy makers.

Emma Briec
Beate Müller

Contents

HELIOS—High Energy Lithium Ion Storage Solutions: Comparative Assessment of 4 Chemistries of Cathode for EV and PHEV Applications

Frédérique Del Corso, Horst Mettlach, Mathieu Morcrette, Uwe Koehler, Cedric Gousset, Christian Sarrazin, Ghislain Binotto, Denis Porcellato and Matthias Vest

Abstract HELIOS is a 4 year project to carry out a comparative assessment of 4 types of lithium-ion battery technology (NCA, LFP, NMC and LMO-NCA or LMO-blend/Graphite). The assessments concern traction batteries for the automotive sector (Electric Vehicles and Plug-in HEV). The evaluations are carried out on 'real' size high energy cells with a capacity of approximately 40 Ah, produced industrially. In total, up to 220 cells have been employed across the various cell types and test activities (safety tests on new and pre-aged cells), cycling and calendar tests (12–15 months). The comparisons have been achieved from

F. Del Corso (✉)
RENAULT SA, 1 av du Golf, API TCR LAB 0 12, 78084 Guyancourt Cedex, France
e-mail: frederique.delcorso@renault.com

H. Mettlach
GM-OPEL, Adam Opel AG, 65423 Russelsheim, Germany
e-mail: horst.mettlach@de.opel.com

M. Morcrette
CNRS-LRCS, Université de Picardie Jules Verne, 33 Rue Saint Leu, 80039 Amiens Cedex, France
e-mail: mathieu.morcrette@u-picardie.fr

U. Koehler
Johnson Controls, Am Leineufer 51, 30419 Hannover, Germany
e-mail: Uwe.Koehler@jci.com

C. Gousset
SAFT, 111-113 Boulevard A. Daney, 33074 Bordeaux Cedex, France
e-mail: cedric.gousset@saftbatteries.com

C. Sarrazin
EDF, Avenue des Renardières – Ecuelles, 77818 Moret sur loing, France
e-mail: christian.sarrazin@edf.fr

E. Briec and B. Müller (eds.), *Electric Vehicle Batteries: Moving from Research towards Innovation*, Lecture Notes in Mobility,
DOI 10.1007/978-3-319-12706-4_1

laboratory testing and other analysis of full sized battery cells in order to determine comparative assessment of Performance, life, cost, recycling and safety characteristics. This paper makes a review of the main results of Helios project.

Keywords High energy cells · EV and PHEV · Li-ion cells

1 Description of Work Performed and Main Results

The expected final results of the project, that will be available in October 2013, at the end of the project, are summarized as follows:

A detailed technical and economical comparison of the four main lithium-ion vehicle traction battery technologies in current manufacture or development was to be done. The 4 types of cathode materials having been selected as the most promising technologies across the world:

Lithium Nickel Cobalt Aluminum (NCA)
Lithium Nickel Manganese Cobalt (NMC)
Lithium Manganese oxide—NCA blend (LMO-NCA or LMO-b)
Lithium Iron Phosphate (LFP)

NCA is the current mainstream manufacturing technology used by SAFT and regarded therefore the base case against which the other 3 technologies are compared.

The whole comparisons have been achieved from laboratory testing and other analysis of full sized battery cells to determine their:

- electrical performance
- cycle life and storage life
- safety under accident or abuse conditions
- volume cost
- capability for recycling of materials

G. Binotto
INERIS, BP n°2, 60550 Verneuil en Halatte, France
e-mail: ghislain.binotto@ineris.fr

D. Porcellato
PSA Peugeot Citroën, 212, boulevard Pelletier, 78955 Carrières-sous-Poissy, France
e-mail: denis.porcellato@mpsa.com

M. Vest
IME, RWTH Aachen University of Technology, Intzestr. 3, 52056 Aachen, Germany
e-mail: mvest@metallurgie.rwth-aachen.de

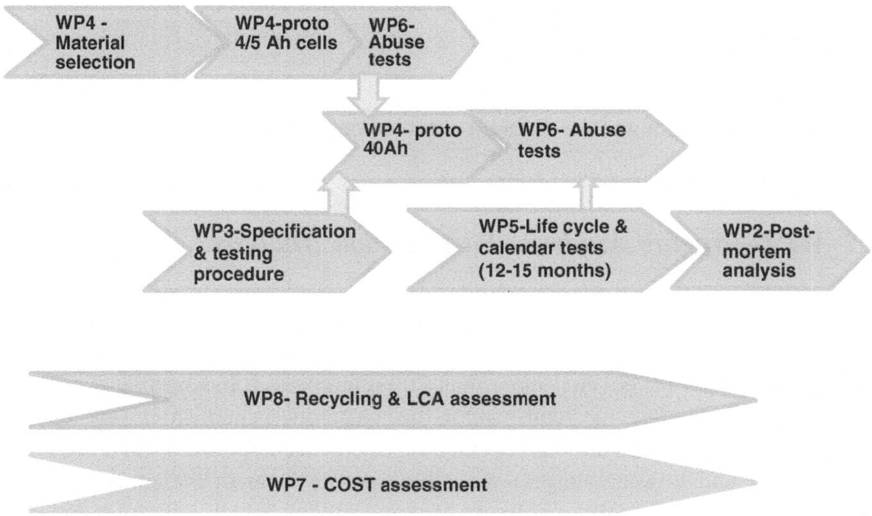

Fig. 1 Technical architecture of Helios project, per work package

In order to carry out the above testing and analysis work it was necessary to develop procedures for each phase. These documents, listed below, will be available for future use of similar activity:

- Cell specifications applicable to both electric and hybrid electric vehicles [1]
- Performance, cycle and ageing test procedures, with links to other existing procedures available world-wide [2]
- Safety test procedures for performance under electrical/thermal/mechanical accident or abuse [3]
- Procedures for producing cost estimates of volume manufacture
- Procedures for handling of used cells and recovery of materials

The majority of the work is performed at cell level, with some module testing (typically 4 cells) being carried out as part of the safety/abuse testing. In all cases, the comparative results will be related to full battery pack size units suitable for complete vehicle installation.

Dedicated Work Packages (WP) visualized in Fig. 1 are focused on the key tasks, namely:

- WP2—Ageing analysis, post mortem analysis
- WP3—Cell specification and test procedures
- WP4—High energy cell manufacture
- WP5—Electrical performance testing (cycling and storage)

- WP6—Safety and abuse testing
- WP7—Economical assessment
- WP8—Recycling assessment

The main approach and the different steps are described in the following section.

1.1 WP2—Ageing Analysis, Post-mortem Analysis

The main objective of WP2 is the post-mortem analysis of the 40 Ah cells produced by SAFT. The first WP2 objective was focused on a full bibliographic review on ageing mechanism [4] covered more than 200 references. It has been analyzed the potential ageing failure mechanisms for the 4 cathodes chemistries (taking into account the operating conditions: cycling capacity, discharge rates, SoC, Temperature, Upper and Lower voltages), than the interface layer of the anode with the electrolyte. Also, to set up the different ageing protocols which will be undertaken in the different partner labs (SEM, XRD, XPS measurements, electrochemical testing…). According to the high number of cells to analyze, a high flow of samples was organized. The cells coming from testing institutes (EDF, CEA, ZSW, AIT, RWTH-ISEA and ENEA) were delivered to SAFT for disassembling the electrodes to be provided to WP2 partners.

The initial characterization of the materials was completed whereas the characterisation of the intermediate electrodes at 45 and 60 °C, as function of chemistry and ageing protocols, is still in progress. This meant a huge amount of work and also remarkable time consuming. Moreover, the difficulties in handling some aged electrodes must be highlighted, because of the bad adhesion of the active material (after cycling) on the current collector (see Fig. 2).

A report with pristine (t = 0), intermediate (t = 6 months) and final (t = 12–15 months) electrochemical and chemical characterization of 40 Ah cells will be issued, first images can be found in Fig. 3.

1.2 WP3—Cell Specification and Test Procedures

The Performance and Aging Test Procedures (Deliverable 3.2) [2] were already streamlined on Helios web site (http://www.helios-eu.org/).

Although, the HELIOS cycle life profiles are based on well-established USABC and ISO standards, there was the question how they correlate to real world driving.

As a next step it is planned to compare the HELIOS test cycles (visualized in Fig. 4) with the real life cycle profiles based on the computer simulation of the battery power profile for HEV/PHEV/EV vehicles.

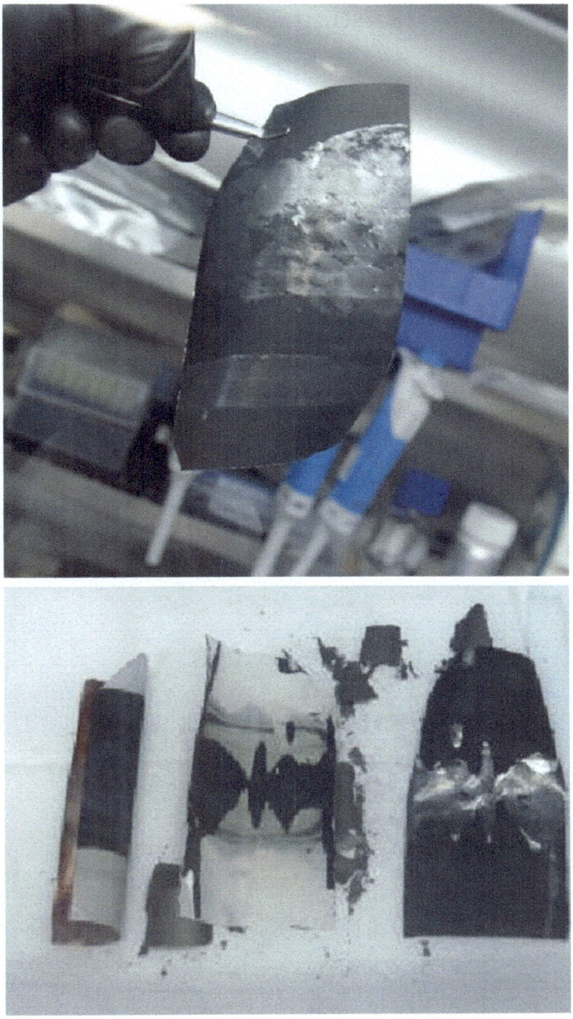

Fig. 2 Photos of electrodes (*left middle*, EV 30 °C sample, *right end*, PHEV 45 °C sample)

It may be possible to calculate a rough estimation of the vehicle mileage based on the number of cycles achieved during the testing in WP5 according to the HELIOS cycle profiles.

The battery system (pack level) specification as described in Deliverable 3.1 [1] of the HELIOS project is described in Table 1.

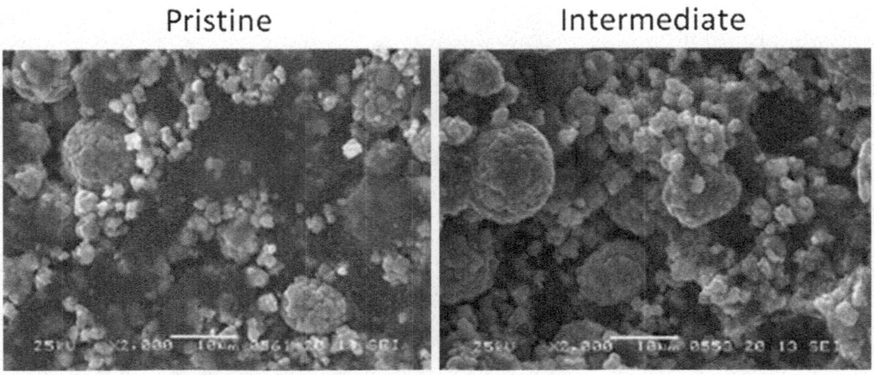

Fig. 3 SEM (ENEA) of pristine (*left panel*) and intermediate (*right*) NMC electrodes

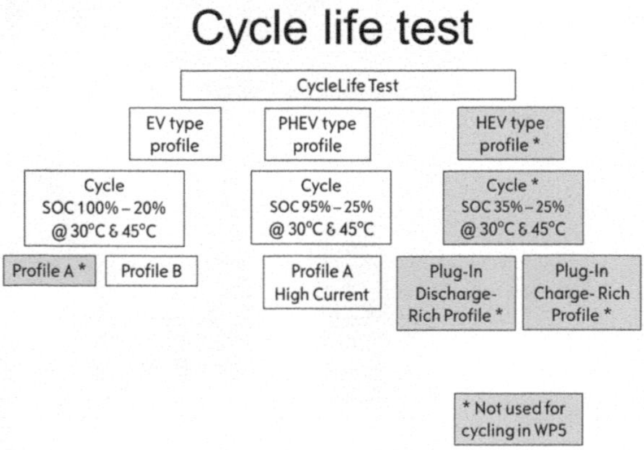

Fig. 4 Overview of applicable test profiles

Table 1 Specification for EV cell (70 and 45 Ah) and PHEV cells (45 Ah)

Type	EV cell	EV cell 45 Ah	PHEV cell
Peak power 15 s (W)			900
Peak power 30 s (W)	1,000	645	850
Peak power 45 s (W)	750	482	
Specific power 30 s (W/kg)	717	717	1,104
Specific energy (Wh/kg)	179	179	150
P/E ratio	4	4	7
Mass (g)	1,400	900	770

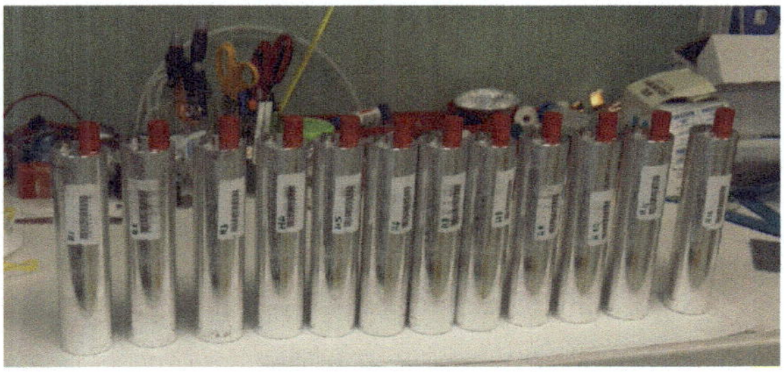

Fig. 5 Picture of 40 Ah cells (NCA chemistry) provided by SAFT and undergoing calendar life tests

1.3 WP4—High Energy Cell Manufacture

Various active materials for positive electrode were investigated and qualified for their electrochemical properties. In a pre-study the behavior of the materials was investigated by the use of small cells with approximately 0.5 Ah. These cells were tested mainly for their safety and aging properties (WP6). As the results were quite promising, the manufacture of large cells was set up by SAFT. They used their industrial lines to produce cathode electrodes and 40 Ah cells (see Fig. 5). As the goal of Helios project is to compare 4 electrochemistries for positive electrode, we've kept the same negative electrode (graphite) and electrolyte used by SAFT in their commercial cells.

Around 60 cells per chemistry were manufactured and delivered to WP5–6 and 8 partners to run life cycle, safety and recycling tests.

1.4 WP5—Electrical Performance Testing (Cycling and Storage)

Concerning the results obtained, cycling tests for the reference chemistry (*NCA*) and two of the alternative chemistries (*NMC and LMO blend*) were started in 2011 and most of these cells have reached more than one thousand EV-cycles and close to two thousand PHEV-cycles. The cycling of LFP cells has started at the beginning of 2012, due to later delivery and problem of process of the electrodes. Concerning calendar storage, the first three chemistries have reached about 12 months calendar storage while LFP cells have reached about 6 months storage, as described in Table 2.

Complete characterization was performed on each single cell: Ragone measurements (*capacity as a function of current rate*), dynamical internal resistance and nominal capacities determined at C rate and reference temperature (30–45 and

Table 2 Summary of the position in cycling and storage for EV, PHEV cycling and calendar life tests, for the Li-ion cells manufactured by SAFT and tested within the Helios project—WP5

	T 30 °C	T 45 °C	T 60 °C
EVs	*Cycles*	*Cycles*	
NCA	1,710	1,400	
NMC	1,380	1,730	
LMO blend	1,910	1,600	
LFP	1,080	770	
PHEVs	*Cycles*	*Cycles*	
NCA	2,300	3,800	
NMC	1,800	2,240	
LMO blend	2,200	3,690	
LFP	1,200	2,170	
Calendar		*Storage*	*Storage*
NCA		18 months	12 months
NMC		12 months	12 months
LMO blend		12 months	12 months
LFP		6 months	6 months

Table 3 Nominal capacity values determined for the Li-ion cells to be tested in WP5

Capacities	NCA	NMC	LMO b	LFP
Ah nominal	41	38	28	35

60 °C). The nominal capacity values, determined by WP5 partners, depending on the chemistry, are reported in Table 3 as follows:

Concerning **EV-cycling**, the State of Health (SOH) is reported in the graph below for the four chemistries at 30 °C (Fig. 6a) and 45 °C (Fig. 6b):

Concerning **PHEV-cycling**, the State of Health (SOH) is reported in the graph below for the four chemistries at 30 °C (Fig. 7):

Then, storage cycling have been performed at 45 and 60 °C on 4 cells per technology (one was removed at 6 months, 2 at the end of the test and the 4th was kept to run abuse test on presaged cells).

The State of Health (SOH) is reported in the Fig. 8 for the 4 chemistries at 60 °C (which is the most severe conditions for accelerating tests).

Main results: NCA and LMO-blend chemistries give the best results at 30 °C for EV profile. But only NCA is best at 45 °C for EV and PHEV applications with the type of cycles used.

In storage conditions, NCA gives also the best results but NMC and LMO-blend are rather good.

We can underline that NCA is indeed a commercial cell so based on optimized formulation, which is not the case for the other chemistries (anodes and electrolyte are the same and samples produced on industrial line but with very short optimization process). So this choice taken at the beginning of the project, to have only

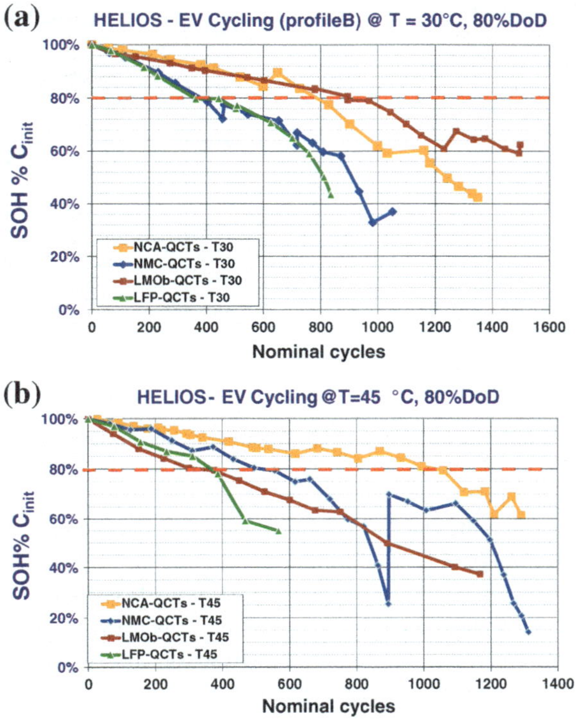

Fig. 6 EV cycling @ 45 °C—comparison of the 4 technologies

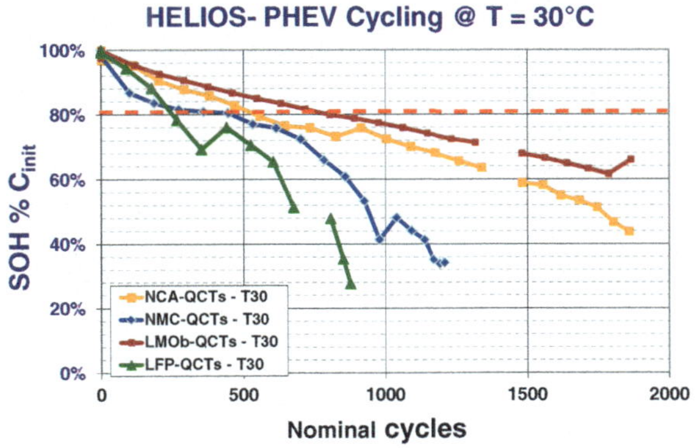

Fig. 7 PHEV cycling @ 45 °C—comparison of the 4 technologies

one modified parameter (cathode composition) influences indeed the final performances.

All the complete results (capacity decrease, impedance and resistance evolution …) will be presented at the end of the project (work under progress).

State of the art: It's difficult to find in the literature, some data about the comparison of the main Li-ion technologies for EV and PHEV applications, on representative cells (25–100 Ah) [5].

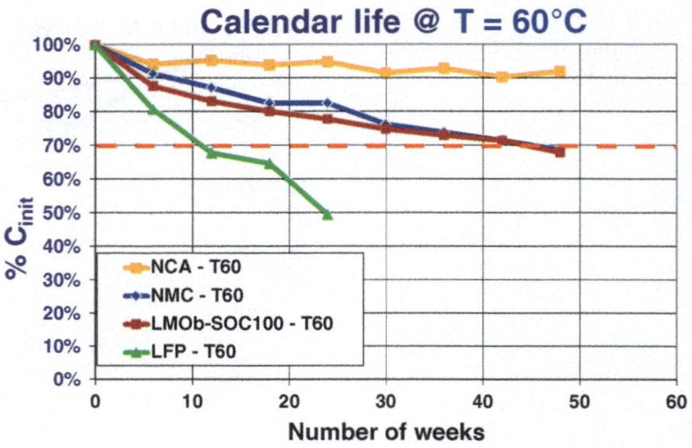

Fig. 8 Calendar life @ 60 °C, 100 % SoC—comparison of the 4 technologies

Netherless, we can find energy and power evolution of Saft VL45 E (for EV application) during DST cycle at 80 % DoD and storage test at 100 % SoC, 40 °C [5].

The cells (NCA/graphite) ensures an excellent calendar life (>1,500 days) and very good stability during cycling (>2,500 cycles), as we can see in Helios project, even if the cells studies are VL41 M cells (which are High energy PHEV design).

Very detailed aging study is described [6] until 450–500 days on NMC Li-ion pouch cell but only at 10 Ah. However, it's very interesting to notice the evolution of the capacity (decreasing) and the resistance (increasing) with the temperature (25–60 °C) and with % SoC (20–100) as we did in Helios project.

We can find also comparison of commercial battery cells (13 different cells from 2 to 70 Ah and for different applications have been full characterized and classified) but there's no data on life performance [7].

At last, proceedings from international conferences, like AABC, presents data and results from OEM's or battery suppliers but most of them are partial for confidential reasons.

So the study performed in Helios project is a complete comparison of the 4 technologies, used in (electric, Hybrid and PHEV) vehicles, which very long testing period.

1.5 WP6—Safety Test

The safety of operation is a key point to allow lithium-ion batteries technology to be widely used for electric vehicles. According to the several types of positive active material dealing in the HELIOS project, they do not have exactly the same performances in terms of specific energy, cycling life time and safety (see Table 3).

The WP6 has established a review on the chemical runaway mechanisms under abuse conditions (in term of safety) [8] to perform the tests and to evaluate these various types of lithium-ion batteries toward electric vehicle applications based on

Fig. 9 BTC (adiabatic calorimeter)

the definition of safety tests procedures provides by WP3 [3], and by using a standard experimental protocol.

The measurement of the reactivity and of the thermal evolution of different positive electrode materials, from the determination of kinetic parameters and approximate enthalpy reactions have given different results depending on the nature of the material, i.e. pristine material or cycled and 'charged' (from a charged cell) material; and allow us to have a better overview through a real comparison of the exothermic reaction on positive electrode material.

The thermal and electrical abuse tests (using Accelerating Rate Calorimeter—ARC or Battery Test calorimeter—BTC, see Fig. 9) performed on forty '0.5 Ah' cells (10 per each technology) have not led to strong thermal runaway or fire and all the selected technologies could be kept to be tested at the large cell level (40 Ah).

The main objective of the WP6 task consists in the evaluation of high energy cells in abuse conditions. The safety tests are performed on full size batteries (28–41 Ah, see Table 3) produced by SAFT in 4 versions of different chemistries, based on thermal, mechanical and electrical tests:

- Controlled crush (100 % SoC, radial and axial positions)
- Nail penetration (100 % SoC, radial position)
- Thermal stability (BTC, 100 % SoC, Begin of Life—BoL and pre-aged)
- Simulated Fuel Fire (BoL, 100 % SoC, axial position)—see Fig. 10
- Elevated temperature Storage (BoL, 2 months storage at 20 % SoC and 50 % SoC, radial position)
- Rapid charge/discharge (BoL, 100 % SoC)
- Thermal Shock Cycling (BoL, from −40 °C to +75 °C, 50 % SoC, radial position)
- Overcharge (BoL, 1C 200 % SoC)
- Short circuit (BoL, 100 % SoC, Rcc = 0.31 m Ohm)
- Overdischarge (BoL, 1C rate from 100 % SoC down to −100 % SoC)

Gas, smoke and flame, released from the batteries tested during abuse tests (crush tests, nail penetration, simulated fuel fire, overcharge) were analyzed

Fig. 10 Simulated fuel fire (100 kW/m^2)

Abuse tests	Level acceptability		Cells tested (ZSW/INERIS)			
	HELIOS	SANDIA	LMO-NCA	NMC	NCA	LFP
Overcharge	3	3	4 *	5	5 → 7	4
Short circuit	2	2	0 *	4	5	4
Thermal Shock cycl	2	2	0 *	0	0	0
Overdischarge	2	2	2 *	2	2	2
Nail penetration	4	3	4 *	5 -6	5	4
Nail penetration (aged cells)			4-5	4	5	4
Crush test (Rad)	3	3	4 *	6	5 → 6	6
Crush test (Axial)	3	3	3 → 4 *	3 → 4	6	3 → 4
Simulated Fuel Fire	5	3	0	0	0	0
Elevated Ture stor.	2	2	0	0	0	0
Thermal Stability	4	3	5 *	5	5	3 → 4
Thermal stability (aged cells@45C)	4	3	6 * 4 *	4 4	6 6	4 4
Module cells (2S2P)	HELIOS	SANDIA	LMO-NCA	NMC	NCA	LFP
Short circuit	2	2	3*	3	3	3

Fig. 11 Abusive tests performed on 40 Ah cells (new and presaged)—synthesis of the results

following the WP3's recommendations (analyzed gases: CO, CO_2, NO_x, HCt, O_2, HF, HCl, HBr, H3P, Aldehydes).

In the Fig. 11, we've detailed all the results of abuse test performed for each cell by ZSW and Ineris on new and presaged (yellow) cells. In red, it's underlined tests have failed, considering Helios level of acceptability which is less severe than SANDIA criterias. LMO-blend cells have lower capacity (28 Ah regarding to 35–41 Ah), so, this chemistry can't be compared to the others.

Considering the whole spectrum of abuse tests performed on 40 Ah large cells (60/76 abuse tests performed), none of the technologies has a fully satisfactory behaviour.

Extrapolation to pack level and preconisation for safety design and chemistry are under progress.

1.6 WP7—Economical Assessment

The active materials and cell components cost have been evaluated by suppliers. The BatPac tool developed by Argonne was used to estimate the cell cost.

To determine the annual quantities of each component, we based the cost estimation on the mass decomposition given by SAFT according to the recipe they used to manufacture, NMC/C, LMO-NCA/C, NCA/C, LFP/C cells but all the cells have not the same capacity (28–41 Ah) due to process or optimization difficulties. So, we have to rework on the design with Batpac tool to get 40 Ah cells for the 4 chemistries, compare all the prices in $/kWh.

At cell level, for PHEV and EV application, the cost estimation is the followingin Fig. 12 ($/kWh).

We can notice that prices decrease of 10 % (even 15 % for LFP chemistry) between 50,000 and 200,000 packs, due a volume effect. NMC is most competitive technology for PHEV, and NMC and NCA are the best one for EV application.

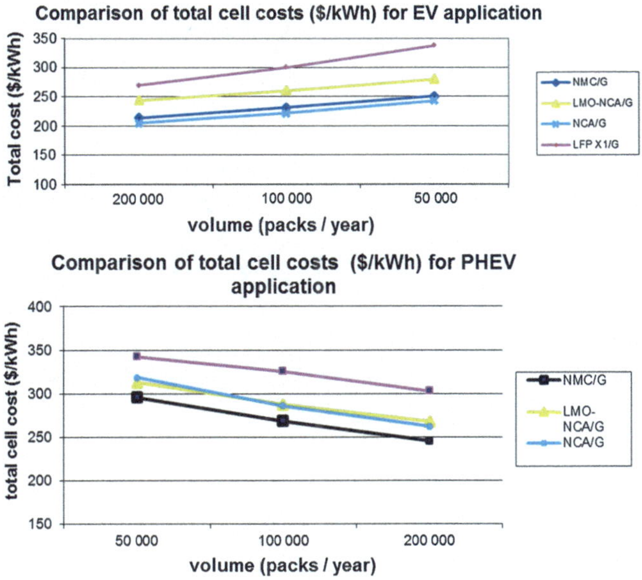

Fig. 12 Comparison of total cell cost for PHEV and EV application for the 4 chemistries

Furthermore, cell cost is higher for PHEV application.

Cost assessment at pack level is under progress taking into account also cooling system, battery management system … for 45 Ah cells (PHEV application) and 70 Ah cells (EV application).

1.7 WP8—Recycling Assessment

The objectives of WP8 are to identify potential recycling processes guided by their technical feasibility and respective possible output products, to validate experimentally and to estimate the environmental impact and costs of the selected recycling concepts for each technology studied (LFP, NMC, LMO-NCA and NCA/C).

A literature research about lithium has been conducted. The economic aspects of lithium, such as reserve, application, demand and price, have been explored and analyzed. The primary (from brine and minerals) is also summarized and discussed and can serve as the advice for secondary production of lithium.

Four potential recycling concepts (visualized in Fig. 13) were identified related to achievable recycling efficiency, productivity, environmental impact, costs and market needs.

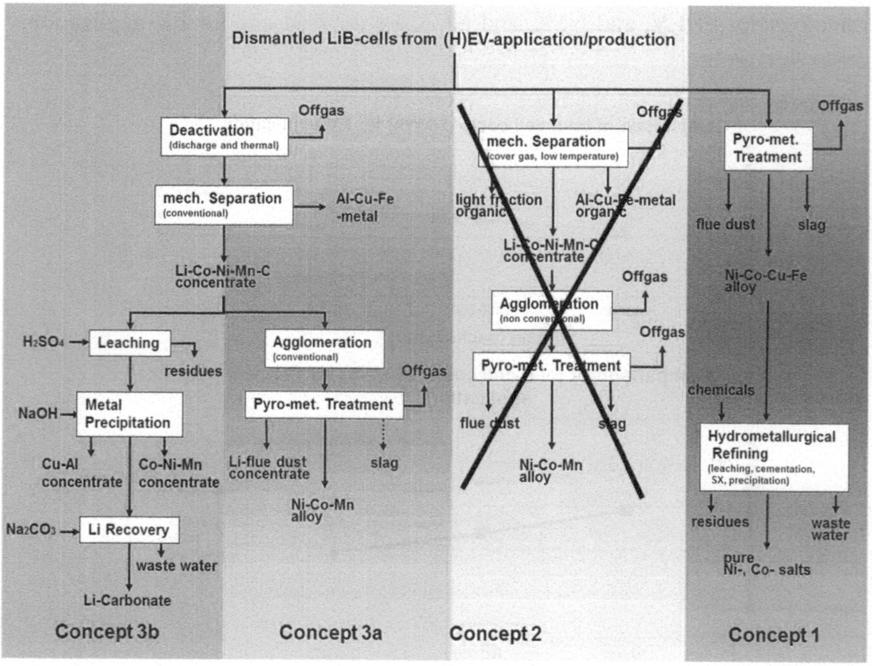

Fig. 13 Schematic overview of the different recycling concepts described in the WP8 (Concept 2 has not been chosen for safety reasons)

In concept 1 (Pyrometallurgy) spent cells of Lithium Ion Battery—(LiB) are directly treated in a furnace at temperatures above 1,500 °C. All organic components of the cells burn or reduce metal oxides like Co, Ni, Fe and Mn. The metallic Al from the casings and the conductor foils burn exothermally and add to the formed slag. Li is mainly slagged. The multi-alloy containing Co, Ni, Mn, Fe and Cu is because of its complexity not sellable and needs therefore further hydrometallurgical treatment. The alloy is leached and each metal is selectively precipitated or via solvent extraction separated. At the end high purity metal salts are gained.

The recycling concepts 3a and 3b start with a permanent deactivation of the LiB cells to assure save handling in the following process steps. During deactivation the cells are pyrolysed at temperatures around 500 C. At those temperatures the volatile components evaporate.

During pyrometallurgical treatment the agglomerated concentrate is heated up to 1,500 °C. At those temperatures the carbon reduces the metal oxides like Co, Ni and Mn, which form a metal alloy. The Li stays in the slag or is reduced and evaporated. The evaporated Li is oxidized in the atmosphere and then collected as flue dust. The slag can be optimized to support the Li evaporation to achieve a valuable Li-oxide flue dust concentrate. The flue dust can be treated by existing hydrometallurgical Li wining processes.

In the flow diagram of 3b, the hydrometallurgical treatment starts with a leaching step of the electrode material with sulfuric acid, followed by a filtration step of the residues. The solution is refined and treated to recovery valuable metals step by step.

In Helios project, we've studied the concept 1 (pyrometallurgical industrial process) and concepts 3a–3b (hydrometallurgical at lab scale)—see Fig. 14.

Recycling trials have been conducted to validate experimentally the recycling efficiency and the chemical composition of the recycling products.

A risk analysis has been performed in respect to safety issues of potential recycling processes and is detailed in Deliverable 8.2.

A report presenting all the extra needs when treating complete large battery packs is in progress.

Fig. 14 Build-up of pyrometallurgical treatment (process 3a)—@ lab scale

2 Conclusion

The project will end in October 2013, all the final results will be widely disseminated.

The potential impact and use is expected to be highly significant. The comparisons cover all of the attributes necessary to support the automotive industry, other research organizations and legislative/funding bodies in their decision making for future electric and hybrid electric vehicles. The socio-economic impact will arise from the guidance the project results will provide in the choice of future battery cell technologies and the way in which cells and batteries can be efficiently and economically employed in use by vehicle owners. In addition the organizations responsible for support to the automotive industry will have a clear view of the most efficient way to direct research and to produce legislation.

The wider societal implications of the project are in the fact that future electric and hybrid electric vehicles will be developed from a stronger knowledge base. This will involve both the vehicle OEMs and the supply industry. In this way much of the uncertainty surrounding the adoption of this new technology will be alleviated giving decision makers a clearer view of the potentially most effective investments in research, development and manufacture. The end result will therefore be a more certain advancement into such vehicles with their ability to assist in longer term benefits for the environment, fuel security and European Union employment.

Helios website: http://www.helios-eu.org/.

Acknowledgments The authors thank the European Union for funding the project HELIOS, which brought the opportunity to carry out this collaborative work.

Also, acknowledgements are directed to all the partners involved into this project:

OEM's (RENAULT, Adam Opel AG, Ford, Volvo, CRF, PSA), other industries (EDF, SAFT, JCHaR, Umicore), Research Institutes (AIT, CEA, CNRS-LRCS, ENEA, ZSW, INERIS), Universities (RWTH ISEA and IME, University of Uppsala).

References

1. Helios Deliverable 3.1, High energy cell target specification. http://www.helios-eu.org/
2. Helios Deliverable 3.2, Initial performance characterisation, cycling and calendar ageing test procedures
3. Helios Deliverable 3.3, Report on recommended safety tests for high energy battery cells
4. Kubiak P, Wolfahrt-Mehrens M, Edström K, Morcrette M, Review on ageing mechanisms of different Li-ion batteries for automotive applications. JPS power D 12:03691
5. Broussely M (SAFT), Pistoia G (2007) Industrial applications of batteries, from cars to aerospace and energy storage. Elsevier, Amsterdam, pp 247–255
6. Kabitz S, Gerschler JB, Ecker M, Yurdagel Y, Emmermacher B, André D, Mitsch T, Sauer DU (2013) Cycle and calendar life study of a graphite/NMC-based Li-ion high energy system Part A: Full cell characterization. J Power Sources 239:572–583

7. Mulder G, Omar N, Pauwels S, Meeus M, Leemans F, Verbruffe B, De Nijs W, Van den Bossche P, Six D, Van Mierlo J (2013) Comparison of commercial battery cells in relation to material properties. Electrochim Acta 87:473–488
8. Helios deliverable 6.1, Review on thermal runaway reaction mechanisms events in batteries

Development of Novel Solid Materials for High Power Li Polymer Batteries (SOMABAT). Recyclability of Components

Leire Zubizarreta, Mayte Gil-Agustí, Marta Garcia, Alfredo Quijano, Alexandre Leonard, Nathalie Job, Roberto Renzoni, Angelique Léonard, Martin Cifrain, Franz Pilcher, Volodymyr Khomenko, Viacheslav Barsukov, Eugenia Fagadar-Cosma, Gheorghe Ilia, Peter Dooley, Omar Ayyad, Pedro Gomez-Romero, Farouk Tedjar, Reiner Weyhe, Karl Vestin, Lars Barkler, Iratxede Meatza, Igor Cantero, Stephane Levasseur and Andrea Rossi

Abstract SOMABAT aims to develop more environmental friendly, safer and better performing high power Li polymer battery by the development of novel breakthrough recyclable solid materials to be used as anode, cathode and solid

---

L. Zubizarreta (✉) · M. Gil-Agustí · M. Garcia · A. Quijano
Instituto Tecnologico de la Energía (ITE), Avenida Juan de la Cierva 24, 46980 Paterna, Valencia, Spain
e-mail: somabat@ite.es

A. Leonard · N. Job · R. Renzoni · A. Léonard
Université de Liège, Place du 20 août, 4000 Liége, Belgium
e-mail: Nathalie.job@ulg.ac.be

M. Cifrain · F. Pilcher
Kompetenzzentr – Das VirtuelleFahrzeug Forschungsgesellschaft mbH, Inffeldgasse 21a, 8010 Graz, Austria
e-mail: martin.cifrain@v2c2.at

V. Khomenko · V. Barsukov
University of Kiev, Nemirovich-Danchenko Str.2, Kiev 01011, Ukraine
e-mail: vkg@ukrpost.ua

E. Fagadar-Cosma · G. Ilia
Institute of Chemistry Timisoara of Romanian Academy, Mihai Viteazu 24, 300223 Timisoara, Romania
e-mail: efagadar@yahoo.com

P. Dooley
Cleancarb, Rue Schmitz 2a, 8190 Kopstal, Luxembourg
e-mail: pdooley@pt.lu

E. Briec and B. Müller (eds.), *Electric Vehicle Batteries: Moving from Research towards Innovation*, Lecture Notes in Mobility, DOI 10.1007/978-3-319-12706-4_2

polymer electrolyte, new alternatives to recycle the different components of the battery and life cycle analysis. This challenge is being achieved by using new low-cost synthesis and processing methods in which it is possible to tailor the different properties of the materials. Development of different novel synthetic and recyclable materials based carbon based hybrid materials, novel $LiFePO_4$ and $LiFeMnPO_4$ based nanocomposite cathode with a conductive polymers or carbons, and highly conductive polymer electrolyte membranes based on fluorinated matrices with nanosized particles and others based on a series of polyphosphates and polyphosphonates polymers respond to the very ambitious challenge of adequate energy density, lifetime and safety. An assessment and test of the potential recyclability and revalorisation of the battery components developed and life-cycle assessment of the cell will allow the development of a more environmental friendly Li-polymer battery in which a 50 % weight of the battery will be recyclable and a reduction of the final cost of the battery up to 150 €/kWh is achievable. The consortium is made up of experts in the field and is complementary in terms of R&D expertise and geographic distribution.

Keywords Lithium · Battery · Polymer · Sustainable · Materials · Solid

O. Ayyad · P. Gomez-Romero
Centro Superior de Investigaciones Científicas, CIN2 (CSIC-ICN) Campus UAB, 08193 Bellaterra, Spain
e-mail: pedro.gomez@cin2.es

F. Tedjar
Recupyl, Rue de la Metallurgie, 38420 Domene, France
e-mail: farouk.tedjar@inpg.fr

R. Weyhe
Accurec, Wiehagen 12–14, 45472 Mulheim, Germany
e-mail: reiner.weyhe@accurec.de

K. Vestin · L. Barkler
Lithium Balance, Baldershoj 26C, 1, 2635 Ishoj, Denmark
e-mail: k.vestin@lithiumbalance.com

I. Meatza · I. Cantero
Cegasa Internacional, Artapadura 11, 01013 Vitoria, Spain
e-mail: imeatza@grupocegasa.com

S. Levasseur
Umicore, Watertorenstraat 33, 2250 Olen, Belgium
e-mail: Stephane.Levasseur@umicore.com

A. Rossi
Atos Origin, Albarracín 25, 28037 Madrid, Spain
e-mail: Andrea.rossi@atosresearch.eu

1 State of the Art

With economical, infrastructural and technological advancements, the world's hunger for energy is ever increasing [1]. Finite fossil-fuel resources, nuclear waste and global warming linked to CO_2 emissions necessitate the rapid development of alternative "green" sources of energy. Electricity generated from renewable resources such as solar and wind power offer great potential to meet these future energy demands; however, the output from sources is intermittent while available electricity is required at any time in our daily lives. These crucial energy supply issues, together with the rapid advance and eagerness from the electric vehicle automotive industry (i.e. Electric vehicles and Hybrid electric vehicles) have combined to make the development of radically improved rechargeable batteries a worldwide imperative. Researchers have thus the responsibility for providing the world with better and more efficient batteries.

The science and technology of lithium batteries have dominated the field of advanced power sources and replaced many other batteries in the market, particularly in the areas of communications, computers, electronics, and more power demanding services such as power tools and transportation. The exponential growth in portable electronic devices such as cellular phones and laptop computers during the past decade has created enormous interest in compact, light-weight batteries offering high energy densities. Also, growing environmental concerns around the globe are driving the development of advanced batteries for electric vehicles.

Lithium-ion batteries are appealing for these applications as they provide higher energy density compared to the other rechargeable battery systems such as lead acid, nickel-cadmium, and nickel-metal hydride batteries [2].

Concerning to their use in electric vehicles, Li ion batteries are expected to be one of the most used energy storage devices used for this purpose in the near future. However, in spite of the several advantage of Li ion technology for its use in hybrid and electric vehicles there are still different technological barriers to overcome, such as the performance of the battery, its life, recyclability, cost and safety.

Research on these issues is multidisciplinary and must involve several themes to gather maximized knowledge and critical mass in a research field where step changes are needed.

Concerning *battery materials* the challenge is to find new low cost cathode (nickel and cobalt oxides are expensive and their prices are exploding) and anode materials which allow high energy density and long-life batteries. Additionally, safety problems related to thermal runaway associated to actual commercial electrolytes should also be solved. One interesting alternative for this is the lithium polymer battery (LPB) which uses a solid polymer electrolyte (SPE). The motivation and advantages for using such a polymeric membrane as the electrolyte component in a lithium cell are: (a) Suppression of dendrite growth; (b) Enhanced endurance to varying electrode volume during cycling; (c) Construction of solid-state rechargeable batteries in which the polymer conforms to the volume changes of both electrodes that occur during charge–discharge cycling; (d) Reduced reactivity with liquid electrolyte; (e) Improved safety; (f) Better shape flexibility and manufacturing integrity [3, 4].

Another aspect that will be looked at is the issue of the *recycling of batteries* at the end of their life cycle and the development of technologies to maximize the recovery of materials, in particular for those of high added-value or presenting high environmental impacts.

For existing or near-to-market types of lithium-based batteries, projects dealing with the *comprehension, modelling and management of degradation drivers and processes* with the aim to extend the calendar and operational life of the cells are also essential.

Finally, the environmental *sustainability* of each developed energy storage technology shall be assessed via life-cycle assessment (LCA) studies.

SOMABAT research focuses on overcoming and improvement of different technological barriers of batteries such as the performance, its life, recyclability, cost and safety for their use in EV.

2 Project Description

SOMABAT aims to develop more environmental friendly, safer and better performing high power Li polymer battery by the development of novel breakthrough recyclable solid materials to be used as anode, cathode and solid polymer electrolyte, new alternatives to recycle the different components of the battery and life cycle analysis (see Fig. 1).

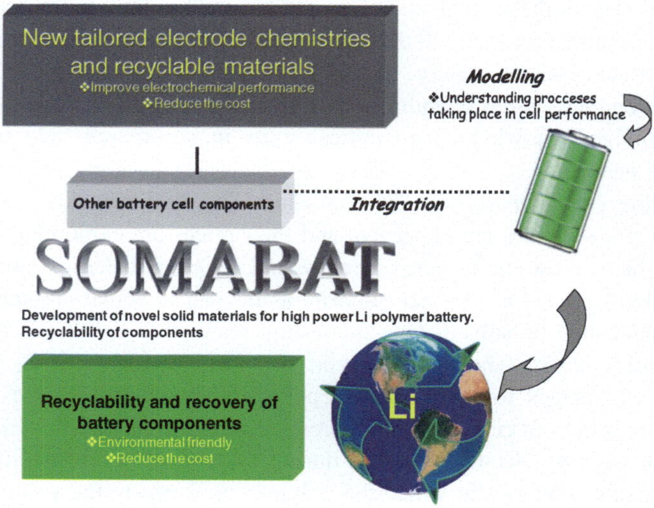

Fig. 1 Schematic representation of SOMABAT project including the main objectives [5]

This challenge is being achieved by using new low-cost synthesis and processing methods in which it is possible to tailor the different properties of the materials. Development of different novel synthetic and recyclable materials based carbon based hybrid materials, novel $LiFePO_4$ and $LiFeMnPO_4$ based nanocomposite cathode with a conductive polymers or carbons, and highly conductive polymer electrolyte membranes based on fluorinated matrices with nanosized particles and others based on a series of polyphosphates and polyphosphonates polymers respond to the very ambitious challenge of adequate energy density, lifetime and safety.

An assessment and test of the potential recyclability and valorization of the battery components developed and LCA of the cell allow the development of a more environmental friendly Li polymer battery in which 50 % weight of the battery will be recyclable.

The general objective of the project is the development of novel breakthrough recyclable solid materials to be used as components (anode, cathode and electrolyte) of a high power and safe Li polymer battery and study and test potential recyclability and sustainability of the battery. The goal is to develop a Li polymer battery with an energy density higher than 220 Wh/kg and a cost lower than 150 €/kWh is the main target.

To achieve the targets novel nanostructured cathode materials based on lithium iron and manganese phosphate will be researched by CIN2 (CSIC-ICN) and UMICORE. The huge advantage of this new material is that it offers maximum energy storage in minimum space, safety and it is environmentally friendly. In addition, anode materials based on synthetic carbon, and other obtained from agricultural wastes will be developed by Université de Liège, Kiev National University of Technologies Design, and ITE. With these materials the energy density will be improved in about 30 % respect to carbon based conventional anodes.

Both electrodes will be much less costly and a lot more reliable than traditional alternatives. Therefore, it will meet the essential requirements for the mass industrial development of electric vehicles.

Moreover, ITE and Institute of Chemistry Timisoara of Romanian Academy will develop new polymeric materials to be used as polymer electrolyte which will reduce outstandingly the safety problems such as leakage, short circuits, overcharge, over-discharge, crush and exposure to fire.

Other strategies which will be followed to reach the ambitious targets are centred on the improvement of materials integration, modelling procedures, and optimizing the management system of the battery. These tasks will be performed by Cegasa International, Virtual Vehicle Competence Center, Lithium Balance, Cleancarb, and Atos.

Expected final results

- Achieve a more environmentally-friendly Li polymer battery in which at least 50 % by average weight of the battery will be recyclable.
- Reduction of the total manufacturing cost of the battery down to 150 €/kW due to recyclability.
- Improvement of the battery safety by the use of solid materials.

3 Project Results up to Now

SOMABAT is a 3 years project which started in January of 2011. In this section, the main results achieved in the project are described divided in the main areas which are under research in the core of the project:

- *Development of synthetic and recyclable material,*
- *Design, development & modelling of a lithium polymer battery,*
- *Recyclability of battery components Sustainability assessment of Li polymer battery.*

3.1 Development of Synthetic and Recyclable Materials

3.1.1 Carbon Anode

In the anode part, the objective is the development of *carbon based hybrid materials composed by graphitisable carbons, novel low-cost synthetic nanostructured carbons and carbon materials obtained using agricultural wastes as precursor as carbon part and metal nanoparticles*, which has the potential to present higher energy density and lower cost than classic carbon anode materials and higher stability and lower environmental impact than lithium metal alloys.

The work has focused on novel carbon/carbon composite material. Such materials were prepared using graphite materials and porous carbon xerogels or carbon material obtained from agricultural waste precursors (i.e. olive stones and orange skin). Composite materials based on different carbon materials have been optimized. It was found that the carbon/carbon composites exhibit high reversible capacity and good cyclability when used as anode materials for rechargeable lithium ion batteries. The graphite based composite with carbon material obtained from agricultural waste precursors with content of 10 wt% exhibits the optimal electrochemical performance with a high reversible capacity over 360 mAh/g. Moreover, the purification of carbon materials in the hydrogen atmosphere at high temperature can further improves the first coulombic efficiency and capacity retention, but decreases the initial capacity of the anode material.

Additionally, the work has dealt with the control of the texture of *carbon xerogels*. In particular, the micropore volume was reduced by addition of secondary carbon precursors either by impregnation or by CVD. In each case, the microporosity was significantly reduced, whereas the mesopore sizes remain unaffected, leaving a good accessibility to the framework. Research on the preparation of ordered mesoporous carbons has allowed identifying some key parameters affecting the structural regularity of the hexagonal mesoporous framework. As can be seen in Fig. 2 introduction of carbon xerogels into the active mass of anode based on

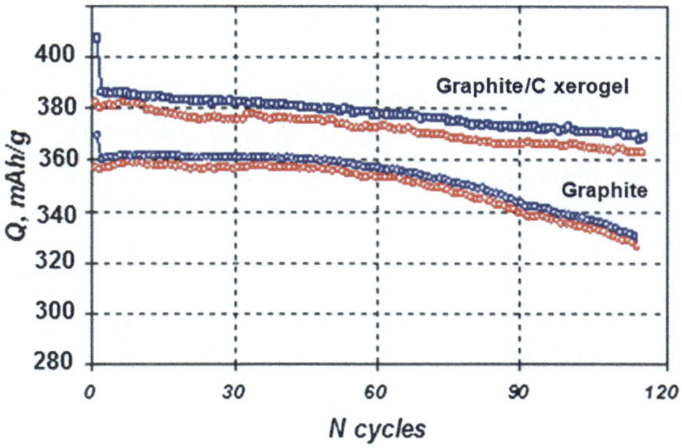

Fig. 2 Specific capacity versus cycle number for anodes based on: graphite SL30; binary mixture

graphite allows to increase a reversible capacity up to 385 mA·h/g per total mass of graphite/carbon content of electrode, and also to improve the stability of characteristics during the cycling.

3.1.2 Polymer Electrolyte

In this case, the objective is the development of safe and highly conductive electrolyte membranes composed fluorinated polymers with nanosized particles and a series of polyphosphates and polyphosphonates.

The effect of several parameters like the effect of amount of different additives such as plasticizers, and lithium salts on the final properties of solid polymer electrolyte has been studied, showing that these two variables affect strongly the final properties of the polymer membranes obtained. After this study the selection of 1st generation polymer membranes was performed. The selected 1st generation polymer membrane presents balanced properties in terms of ionic conductivity, thermal and mechanical properties and was scaled up for integration in 1st generation Li polymer cell. In parallel, the 2nd generation polymer membranes have been under study. The aim of 2nd generation polymer membrane development was to test alternative more stable and environmentally friendly Li salt and plasticizer for their use in polymer membrane composition. The preliminary results show that the alternative plasticizers tested are thermally and electrochemically more stable than traditional carbonates and maintaining the ionic conductivity of the polymer membranes developed.

Additionally, new formulations for polymers and copolymers syntheses with the purpose of improving the electrochemical and mechanical characteristics of the polyphosphate based membranes to fulfil the initial requirements have been studied. The main performed activities were:

Synthesis of phosphate copolyethers from phosphorus oxychloride and polyethylene glycols (PEG 6000, PEG 2000) and membranes based on (co)polyphosphoesters and commercial acrylates, containing lithium trifluoromethanesulfonate by UV curing and their characterisation by FT-IR Spectroscopy, Thermal Analysis

(TG, DSC), EIS, resistance and transference number have been performed. This type of membrane had good conductivity, relatively good mechanical properties but worse stability. New polyphosphoesters based on phosphorus oxychloride, 1,4-butandiol monoacrylate and PEG 2000–4000 were synthesized. Polyphosphoesters–diacrylate-lithium perchlorate composites were obtained by UV curing and characterized by FT-IR spectroscopy and Thermal Analysis. EIS spectroscopy showed good conductivity (3×10^{-5} S/cm) but also low electrochemical stability.

Supplementary work was carried out for searching the best formulation for improving electrochemical and elastic properties required by integration: more sticky membranes to have better adherence to electrodes were developed. Polyphosphates starting from PEG, phosphorus oxychloride and butanediol monoacrylate have been synthesised, from which 90 sticky membranes with dimensions of 86 × 168 mm were obtained. These membranes were sent, in order to develop first generation battery.

After generation cells evaluation, different alternatives were tested by using other polyols instead of PEG-s, namely Polypropylene glycols (for example PPG 1500) in order to obtain polyphosphoesters with improved characteristics such as: crystallinity, conductivity, mechanical properties. Besides, esters of phosphorus derivatives and 1,4-butandiol monoacrylate were obtained, in order to use them as co-monomers or crosslinkers. Membranes obtained by UV-curing or thermal polymerization have been realized. Characterization of these compounds is in progress for the selection of 2nd generation materials.

3.1.3 Cathode

Concerning the design of 1st generation (Gen#1) lithium iron phosphate (LFP) products, the flowsheet was optimized to be able to reach the cost targets of the project. In the same time, performances criteria such as cycle life and power were looked at in order to propose a product suitable for large cells market. Upscaling was done in order to produce large quantities to build scale 1 cells. Products were also sampled to external customers for further market adoption. Then, a high voltage cathode material was successfully developed at lab scale, the benefit of it being to increase the cell voltage and energy density. Substitution of iron by other elements in the LFP structure leads to some inherent loss of power performances. These losses have to be reduced before being able to upscale the 2nd generation (Gen#2) phosphate materials to pilot production.

Additionally, the optimization of two selected synthetic methods for $LiFePO_4$ materials and the corresponding cathodes, namely (i) solvothermal (ii) reflux has been developed, the latter method has been targeted for upscaling. This optimization has included a wide battery of experiments for the fine-tuning of synthesis parameters and a correlation of those parameters with the micro-meso-structure of the $LiFePO_4$ materials obtained. As part of this systematic optimization materials with suitable nanosized particles self-assembled into larger micron-sized aggregates were prepared, a hierarchical structure which provides ideal microstructures for electrode applications. Presently, the work is centered in the characterization of 2nd generation materials and coating of these new materials with carbon and with conducting polymers.

3.2 Design, Development and Modelling of a Lithium Polymer Battery

3.2.1 Integration

The 1st generation design for the SOMABAT Li polymer cells has been performed, with stacked electrode/membrane pouch cell design and large size automotive format, so the scalability of the materials can be proved (see Fig. 3). Since the active area of the electrodes to stack is ~approx. 80 × 160 mm, the slurry formulation and fabrication of the electrodes from the active materials has been developed, while 70 membranes (85 × 168 mm; 50 μm thickness) were prepared required for one cell.

After slurry formulation scale-up and optimisation, several meters of cathode (LFP#1) were prepared. Corresponding quantity of anode using commercial graphite and standard formulation has also been produced for the first proof-of-concept cells assembled. With the available membranes, one cell (expected capacity 2–3.5 Ah) with each selected polymer electrolyte was prepared. This first assembly has revealed swelling issues with polyphosphoester based membranes and difficult handling of fluorinated based samples leading to short-circuit that are being solved for 2nd generation development. Coin cells have been assembled with these materials. In Fig. 4 charge-discharge profile of the one of the 1st generation cell can be seen. Testing results in this format have shown that 1st generation membranes (despite their conductivity being lower than the standard liquid electrolyte) can withstand up to 1 C discharge rate at room temperature with stable cycling for more than 400 cycles.

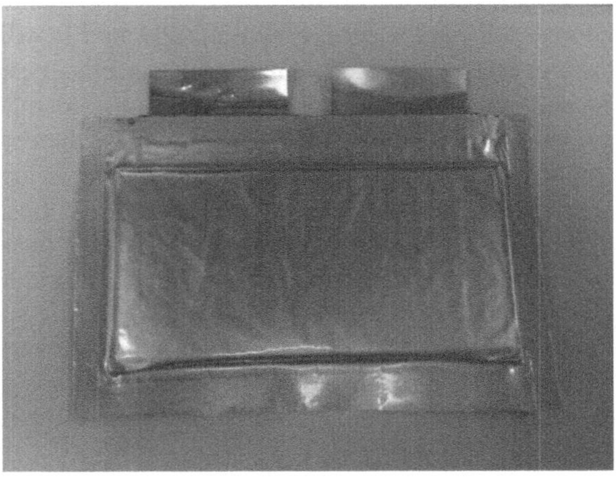

Fig. 3 The 1st generation prototype cell

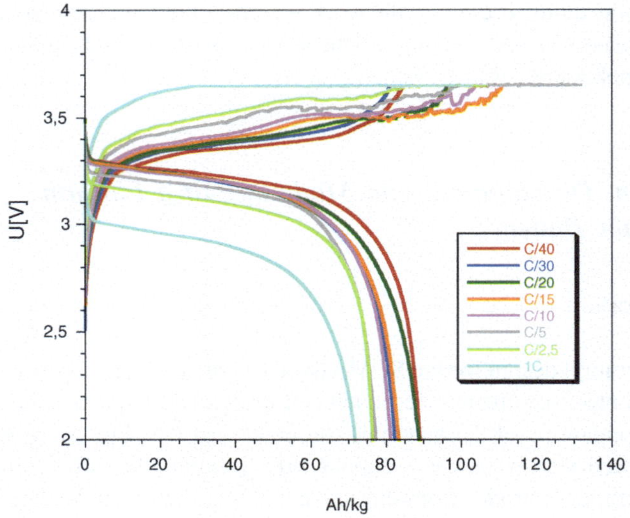

Fig. 4 Charge and discharge profiles of 1st generation cell

Additionally, anode formulation optimisation and scale-up from the C/C composite has been performed. An optimised formulation has been achieved during this period and is ready to be used to complete 1st generation characterisation by assembling smaller area stacked pouch cells.

The concept design of the battery pack with at least 4 cells connected in series has been presented during the workshop at Timisoara in July 2012. Once the generation 2 materials have been developed and tested a second set of pouch cells will be manufactured.

3.2.2 Modelling

In the approach of modelling the SOMABAT battery a multi-scale model of the lithium-polymer battery is proposed. The models will be used to give an insight to the proposed battery module regarding temperature and electrical distribution and to support cell and module optimization. In the attempt of building up a multi-scale model of the lithium-polymer battery with full numerical integration, the following length scales (model levels), are distinguished: device level ($\sim 10^{-1}$ m), electrode level ($\sim 10^{-4}$ m) and particle level ($\sim 10^{-7}$ m). Besides in length scale, they also differ in the time scales of the physical effects of interest. With inputs regarding the geometries of cells and module, the model for the highest level of geometry was implemented. Also, the core set of governing equations of the electrochemical model for the lowest level (particle level) and the interface conditions were defined, the homogenization procedure was set up and implemented in PYTHON. Thereby, the finite element method (FEM) has been chosen on all levels, because of its flexibility. On the top level, the 3D FEM toolbox ELMER is used. The solver, which was coded for ELMER in FORTRAN, was updated and speeded up by some changes in the technical approach to the solution process.

3.2.3 Battery Management System

The environmental testing and firmware development for the *battery management system hardware* that was developed and prototyped during the first year of the project has also been performed. This includes software design, source code development, reviews, validation and testing the system for stability on terms of electromagnetic immunity, vibration robustness and temperature operating range. In Fig. 5 detailed view of CMU diagnostics software is shown.

At the end of the second year prototypes for a battery management system suitable for testing the next generation of SOMABAT battery cells have been completed.

3.3 Recyclability of Battery Components

The recycling work is divided in 3 periods, mainly international comparison of existing recycling processes, investigation of SOMABAT battery materials in viewpoint of recycling procedures, and development of two alternative recycling

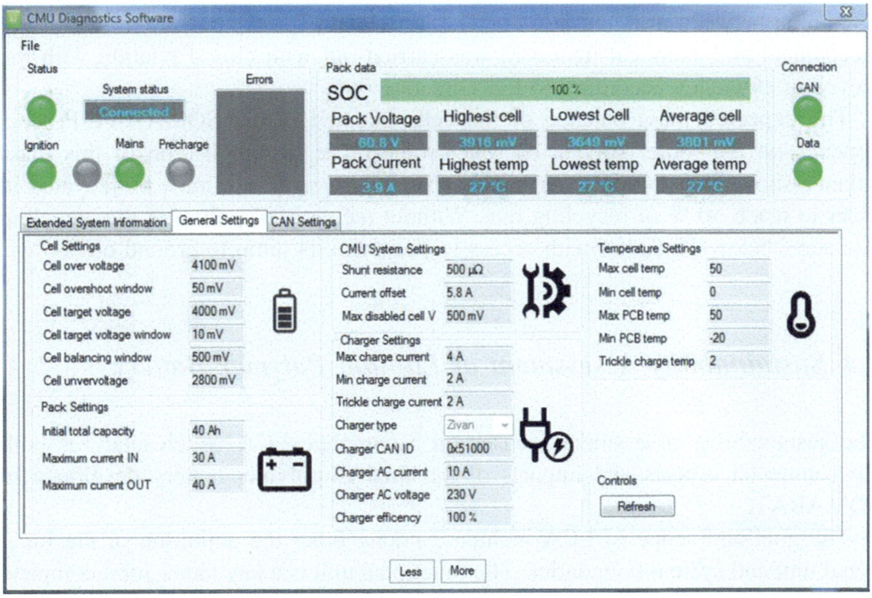

Fig. 5 Detailed view of CMU diagnostics software

processes for this newly developed SOMABAT battery. This development foresees a basic design of hydro- on the one hand, and pyrometallurgical process on the other hand. This process design has to be verified by installing and testing this recycling technique at laboratory scale.

Since beginning of SOMABAT project, there has been activity on the following tasks:

Legislative requirements on transportation, packaging and recycling of end-of-life (EOL) Li-ion Batteries, investigation of theoretically possible recycling routes, data consolidation on existing Li-Ion battery recycling processes (one dedicated, and one non-dedicated recycling facility), evaluation and comparison of these processes in terms of recycling efficiency of recycled materials, compliance with legislative requirements, environmental impact of process, economic performance, flexibility on varying input-materials due to changing Li-ion subtype compositions.

After comparison of technical process performance, the economic features of possible process routes, pilot scale plants as well as industrial implemented plants have been figured out. Summarizing these results, and gathering all detail data of project members on battery components, a specific SOMABAT recycling process to combine economic efficiency with technical optimisation of recovered metals has been designed.

Additionally, a mechanical treatment to achieve the safe and efficient access to active material, physical sorting to separate between metals, oxides and polymers have been developed. The mechanical trials were finalized with success as a separation was done by up to 90 % of each fraction reported. The chemical treatment was made a safe way at room temperature and iron-oxide was recovered from iron based cathodes using dissolution/precipitation shuttle process leading to efficient separation between iron and lithium. The last metal is precipitated as Li_2CO_3. The mechanical and chemical treatment was carried out with closed relation with the recycling efficiency according to EU Directive 066.

The expected mass balance of the cells planned in the SOMABAT Process depends on the route (solvent or water route). The decomposition of this mass balance shows that the polymer has a substantial weight and must be recycled in order to reach 50 % of recycling rate. Without recovery of polymers the recycling rate stays below 50 % but with access to polymers its jump to around 60 %.

3.4 Sustainability Assessment of Lithium Polymer Battery

The sustainability assessment focuses on a complete LCA which analyzes both environmental aspects and impacts of the final Li-polymer battery developed in SOMABAT.

The goal and scope of LCA includes, among other the definition of the functional unit and system boundaries. The functional unit is a key factor for a complete

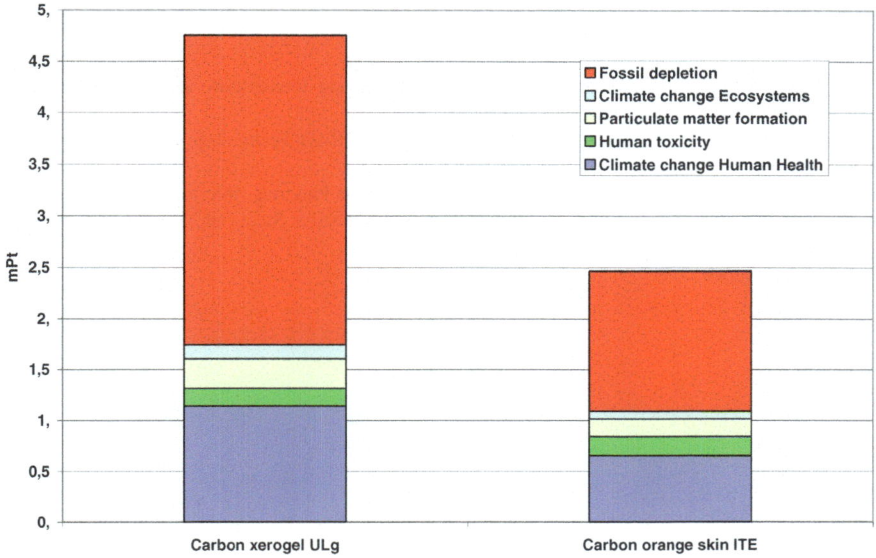

Fig. 6 Comparison of impacts of carbon xerogel and carbon material obtained from orange skin, method: ReCiPe endpoint (H)/World ReCiPe H/H/Single score

evaluation of LCA and, in this case, it is a given amount of energy (30 kWh) accumulated by the battery and then delivered to the powertrain for an electric vehicle capable of sustaining 4,000 charge cycles at 80 % maximum discharge giving at least a 210.000 km operation during the vehicle design life time. System boundaries comprise the entire life cycle of the battery from “cradle to grave”.

During this period, analysis and modelling of several materials were performed. Data were obtained from different partners and preliminary analysis on parts of the battery was carried out. Carbon xerogels obtained by ULg and orange skin carbon from ITE were analysed and compared (see Fig. 6).

The same approach was applied to membranes. Polyphosphonate and PVdF-HFP based membranes were analysed and the comparison was performed.

Acknowledgments The research leading to these results has received funding from the European Community’s Seventh Framework Programme (FP7/2007–2013) under Grant Agreement n°266090 (SOMABAT).

References

1. Man ME, Bradley RS, Hughes MK (1998) Global-scale temperature patterns and climate forcing over the past six centuries. Nature 392:779
2. Lithium Battery Energy Storage (LIBES) Publication (1994) Technological Research Association, Tokyo
3. Megahed S, Scrosati B (1995) Rechargeable nonaqueous batteries. Interface 4(4):34–37
4. Brummer SB, Koch VR (1980) In: Murphy DW, Broadhead J, Steels BCH (eds). Materials for advanced batteries. Plenum, New York
5. www.somabat.eu

AUTOSUPERCAP: Development of High Energy and High Power Density Supercapacitor Cells

Constantina Lekakou, Aldo Sorniotti, Chunhong Lei, Foivos Markoulidis, Peter C. Wilson, Alberto Santucci, Steve Tennison, Negar Amini, Christos Trapalis, Gianfranco Carotenuto, Sofie Khalil, Brunetto Martorana, Irene Cannavaro, Michele Gosso, John Perry, Craig Hoy, Marcel Weil, Hanna Dura and Fabio Viotto

Abstract The study focuses on the materials and small supercapacitor cells manufactured in the first period of AUTOSUPERCAP project. The supercapacitor cells presented in this paper are of the type of symmetrical, electrochemical double layer capacitor (EDLC) cells with organic electrolyte $TEABF_4$ dissolved in propylene carbonate (PC) or acetonitrile (AN). Different active electrode materials have been investigated, including novel activated carbon, graphene and carbon nanotubes produced in this project, as well as combinations of these materials.

C. Lekakou (✉) · A. Sorniotti · C. Lei · F. Markoulidis · P.C. Wilson · A. Santucci
Department of Mechanical Engineering Sciences, University of Surrey,
Guildford, Surrey GU2 7XH, UK
e-mail: C.Lekakou@surrey.ac.uk

S. Tennison · N. Amini
MAST Carbon International Ltd, Basingstoke, Hampshire RG22 4BA, UK
e-mail: steve.tennison@MastCarbon.co.uk

C. Trapalis
National Centre for Scientific Research "Demokritos", Agia Paraskevi Attikis,
P.O.Box 60228, 153 10 Athens, Greece
e-mail: trapalis@ims.demokritos.gr

G. Carotenuto
Institute of Composite and Biomedical Materials, National Research Council,
Piazzale Tecchio. 80, 80125 Naples, Italy
e-mail: giancaro@unina.it

S. Khalil
Bayer Technology Services GmbH, Technology Development, CHEMPARK,
51368 Leverkusen, Germany
e-mail: C.Lekakou@surrey.ac.uk

B. Martorana · I. Cannavaro · M. Gosso
New Materials Scouting & Nanomaterials Department, Centro Ricerche Fiat S.C.p.A,
10043 Orbassano, TO, Italy
e-mail: brunetto.martorana@crf.it

E. Briec and B. Müller (eds.), *Electric Vehicle Batteries: Moving from Research towards Innovation*, Lecture Notes in Mobility,
DOI 10.1007/978-3-319-12706-4_3

Supercapacitor cells of 2–4 cm^2 area were fabricated and tested in impedance spectroscopy, cyclic voltammetry and charge-discharge tests. Ragone plots of energy density against power density were constructed from the charge-discharge test data at different current densities. Furthermore, the results of a cost analysis are presented for the main types of supercapacitors investigated.

Keywords Supercapacitors · Activated carbon · Carbon nanotubes · Graphene · Electric vehicle

1 Introduction

Supercapacitors are of interest in electric vehicle (EV) applications as they offer high power density, which makes them most suitable for start up applications and recovery of the braking energy during the many stop-start ups of city driving or driving in traffic jams. In this case, the high power and high current demands place a serious strain on the battery and accelerate its failure. Furthermore, supercapacitors offer the advantage of huge number of cycles and a long cycling lifetime in contrast with the short cycling lifetime of batteries, which means that the use of supercapacitors in the stop-start applications during city driving will extend the lifetime of the battery, which is critical for EV applications. If the energy density of supercapacitors could be also increased more than the current levels of commercial supercapacitors (around 5 Wh/kg), they may be able to be considered in a more energy storage auxiliary role with the battery in the power system of EVs.

I. Cannavaro
e-mail: brunetto.martorana@crf.it

M. Gosso
e-mail: brunetto.martorana@crf.it

J. Perry · C. Hoy
AGM Batteries Ltd, Thurso, Caithness KW14 7XW, UK
e-mail: john.perry@abslpower.com

C. Hoy
e-mail: john.perry@abslpower.com

M. Weil · H. Dura
Institute for Technology Assessment and Systems Analysis, Karlsruhe Institute of Technology (KIT), 76133 Karlsruhe, Germany
e-mail: marcel.weil@kit.edu

H. Dura
e-mail: marcel.weil@kit.edu

F. Viotto
Oerlikon Graziano, 10098 Rivoli, TO, Italy
e-mail: fabio.viotto@oerlikon.com

EV applications may be associated with voltages of 15–48 V for start up applications and also 300–400 V if they have a more serious auxiliary role in parallel with the battery. For this reason, organic electrolytes are considered in this study as they have demonstrated [1] maximum operating voltage to 3 V against 1.1 V for aqueous electrolytes. We have also investigated ionic liquid electrolytes, such as $EMIBF_4$ which could reach a higher maximum voltage of 5 V; however, these electrolytes showed poor power density, as they have high viscosity, slowing down the charge transport at high currents and rates. This study includes work from the first phase of the AUTOSUPERCAP project in which laboratory-scale cells of 2–4 cm^2 were fabricated and tested. In the following sections, the main materials and experimental techniques will be presented, as well as the results of EDLC cell testing and cost analysis.

2 Materials and Experimental Techniques

Symmetric EDLC cells were fabricated comprising the following components: an outer current collector on each side of the cell, usually aluminium foil; active electrode carbon material; a porous separator sandwiched in the middle. The cells included in this paper were impregnated with an organic electrolyte, $TEABF_4$/PC or $TEABF_4$/AN solution. The following carbonaceous materials have been tested as active electrode materials: activated carbon (AC) powder (Fig. 1a) on its own or mixed with conductive additives, such as carbon black or multiwall carbon nanotubes; multiwall carbon nanotubes (MWNTs) on their own or in blends with AC powder at different compositions (Fig. 1b); graphene (Fig. 1c); activated carbon (AC) fabrics (Fig. 1d); sintered activated carbon electrodes on external foil current collectors (Fig. 1e).

For the active electrode materials starting in the form of particulates, these particulates were dispersed in a solution of PVDF in NMP at a ratio of 5 wt% PVDF and 95 wt% carbonaceous active electrode material. A paste was formed from that mixture which was coated to aluminium foil, used as current collector (Fig. 2).

In the first phase of AUTOSUPERCAP project, small symmetric EDLC cells were fabricated and tested of dimensions of 2 or 4 cm^2 for the AC-fabric based cells (Fig. 3).

The cells were tested (Fig. 4) in impedance spectroscopy tests (Nyquist impedance plots) in the frequency range of 1 MHz–10 mHz, cyclic voltammetry in the range of 0–3 V, and galvanostatic charge–discharge tests at different current densities. From the latter charge–discharge test data, Ragone plots were derived from which the energy density and power density of each type of cell were determined.

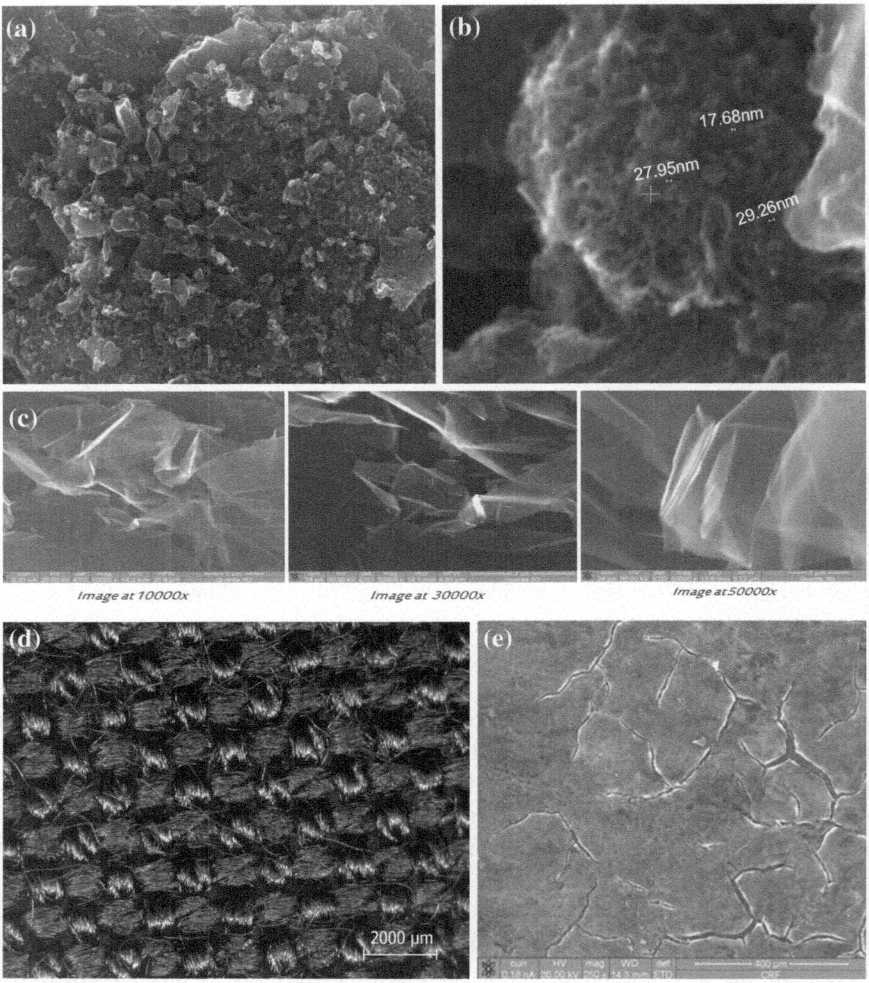

Fig. 1 Starting materials used in the fabrication of active electrodes: **a** AC powder; **b** AC and MWNT coating with 5 wt% PVDF binder, MWNTs produced by BTS, coating fabricated by University of Surrey and SEM photograph by CRF; **c** graphene from NCSR-Demokritos **d** AC fabric by MAST Carbon; **e** sintered phenolic-derived activated carbon electrodes by MAST Carbon

3 Results and Discussion

3.1 Graphene-Based Supercapacitor Cells

Graphene nanoplatelets were produced in different forms and were also functionalised [2–4]. Although their production and treatment did not always yield graphene of the same consistent quality and BET (specific surface area according to the

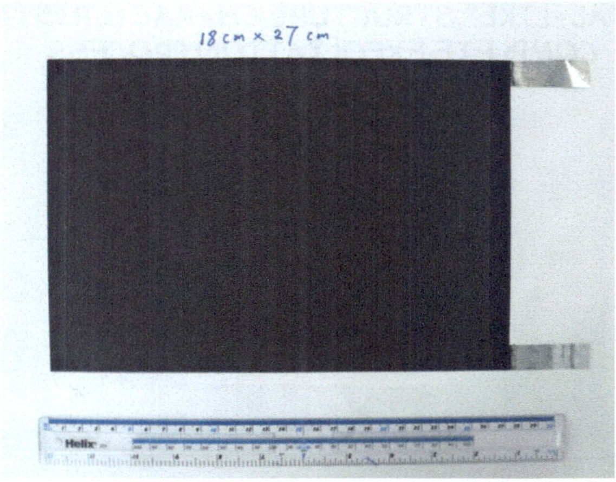

Fig. 2 Active electrode coating on aluminium foil used as current collector, as produced by the University of Surrey

Fig. 3 Small supercapacitor cell testing at the University of Surrey

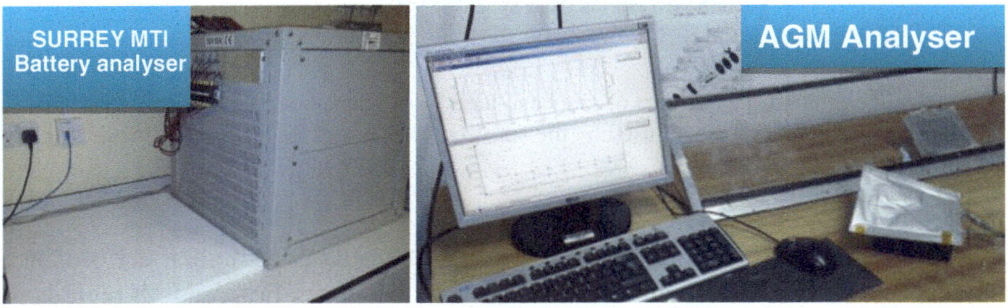

Fig. 4 Some of the supercapacitor testing equipment at the University of Surrey and at AGM Batteries Ltd

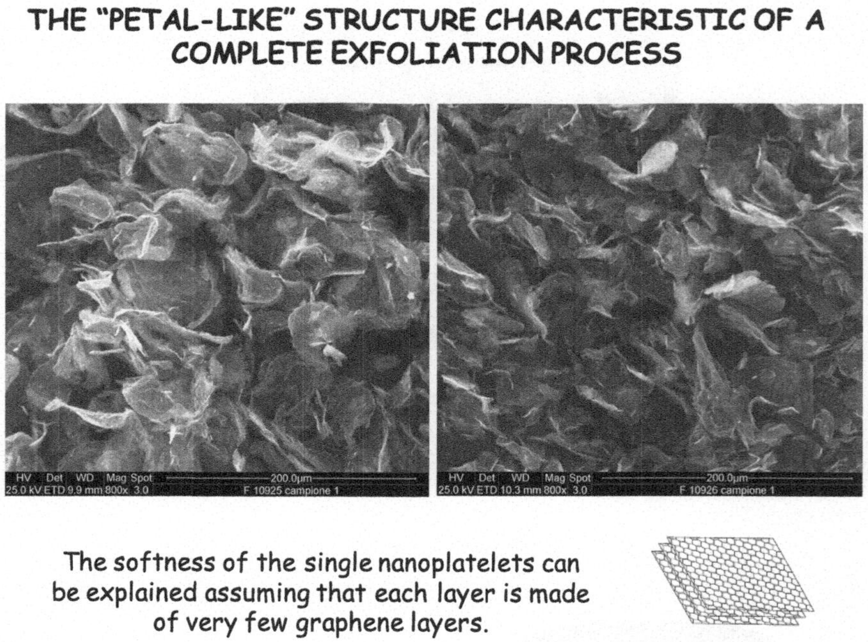

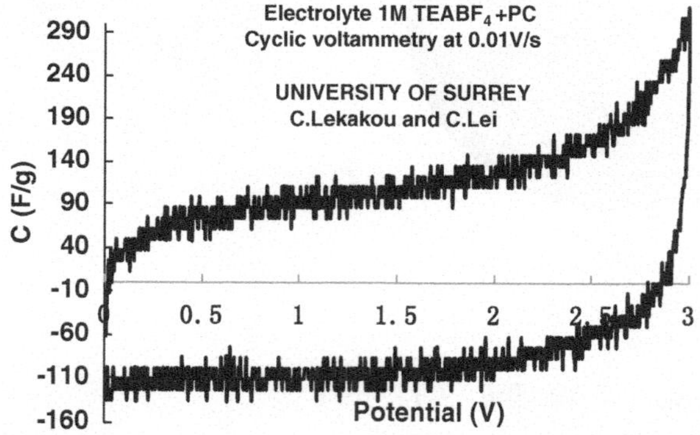

Fig. 5 Cyclic voltammetry test data of graphene-based supercapacitor cells, produced by IMCB-CNR and NCSR-Demokritos, and used by the University of Surrey for further coating processing and cell fabrication and testing

Brunauer–Emmett–Teller theory), some samples resulted in higher performance supercapacitor cells with organic electrolyte as is shown in Fig. 5. With regards to the application of these graphene-based supercapacitor cells to electric vehicle (EV) applications, we believe that further research developments are needed to (a) obtain large quantities of graphene, functionalized graphene and decorated graphene with homogeneous, consistent ad repeatable properties; (b) to achieve graphene coatings of considerable areal mass (above 2 mg/cm^2) so that they can be considered realistically for EV applications: at present within AUTOSUPERCAP, we have fabricated high performance supercapacitor cells with small areal mass graphene electrodes, which make such cells most suitable for electronic applications.

3.2 *Supercapacitor Cells Based on Activated Carbon Coatings and Activated Carbon-MWNT Coatings*

Electrodes with sintered electrode coatings (Fig. 1e) were used to fabricate small cells which demonstrated reasonable supercapacitor performance but cracks in the electrode coatings made them unsuitable for producing large rolls of this material.

First of all, phenolic-derived AC powder-based supercapacitor cells were fabricated and tested for different degrees of burn off during carbon activation. Figure 6 presents the results for cells with 1.5 M SBP-BF4/PC electrolyte. The supercapacitor cell exhibits a most impressive energy density but relatively medium power density. When acetonitrile (AN) is used as solvent in the organic electrolyte, Fig. 7 shows that it is possible to reach higher power density for the phenolic-derived AC coatings, to 10 kW/kg (and possibly higher if we had conducted tests at higher currents). At the same time, activated carbon powder from natural resources has also been tried for which supercapacitor cells could reach higher power density, 18 kW/kg, but lower energy density, 18 Wh/kg, as is presented in Fig. 8. Addition of MWNTs produced by BTS raised both power and energy density performance to a maximum tested 40 kW/kg and 27 Wh/kg, respectively.

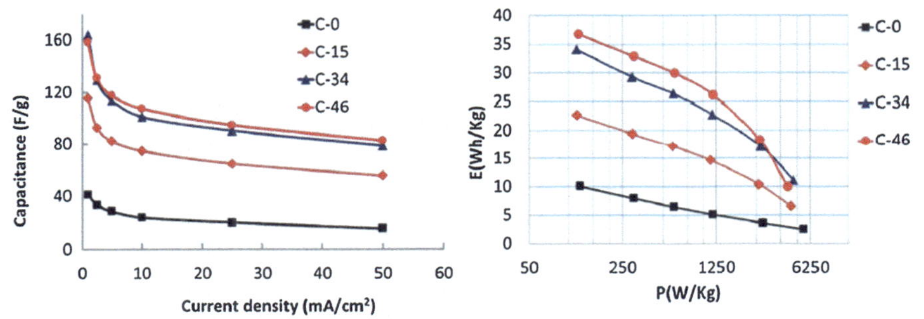

Fig. 6 Graphs from charge–discharge data for supercapacitor cells based on active electrode coatings consisting of AC, 5 wt% carbon black and 5 wt% PVDF [5], where the AC powder is phenolic-derived, produced, carbonised and activated by MAST Carbon to different degrees of burn off (0, 15, 34 and 46 %)

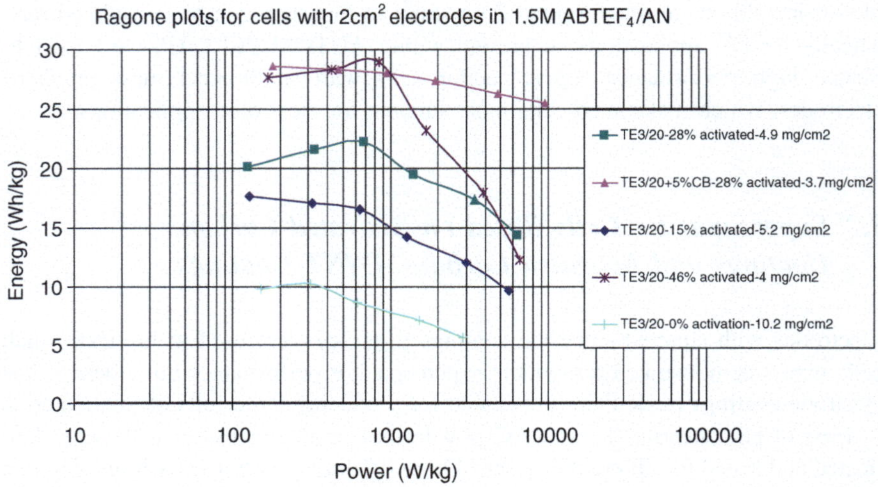

Fig. 7 Ragone plots from charge–discharge data for supercapacitor cells based on active electrode coatings consisting of AC, 5 wt% carbon black and 5 wt% PVDF, where the AC powder is phenolic-derived, produced, carbonised and activated by MAST Carbon to different degrees of burn off (0, 15, 34 and 46 %); electrolyte is 1.5 TEABF$_4$/AN

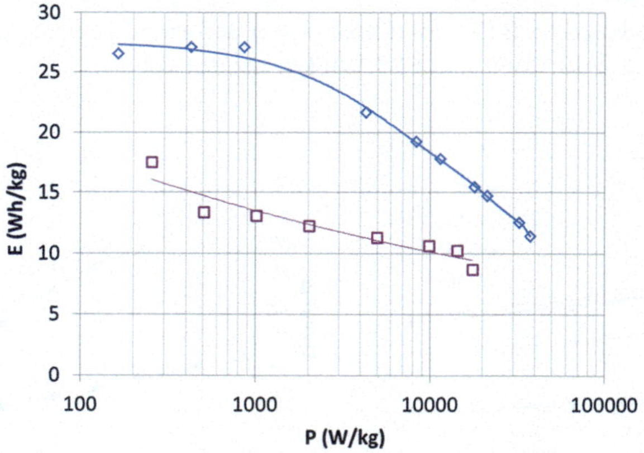

Fig. 8 Ragone plots from charge-discharge data for supercapacitor cells based on active electrode coatings consisting of AC and 5 wt% PVDF (data points: *purple squares*), where the AC powder is derived from natural sources, and also AC, MWNT (Baytubes®) and 5 wt% PVDF (data points: *blue diamonds*); electrolyte is 1.5 TEABF$_4$/AN

3.3 Supercapacitor Cells Based on Activated Carbon Fabrics

Activated carbon (AC) fabrics used as active electrodes have the advantage of avoiding the use of a binder, which is used in coatings, leading to higher conductivity and surface area. Tests so far have shown a high energy density over 40 Wh/kg and a maximum tested power density of 10 kW/kg.

3.4 Life Cycle and Cost Analysis

A Life Cycle Assessment (LCA) and a Life Cycle Costing (LCC) were carried out for the novel supercapacitors, in parallel with the research developments in materials and manufacturing methods. An initial analysis of the environmental burden of nanoenabled supercapacitors has been published [6]. Initial cost calculations were carried out for different types of supercapacitor cells in AUTOSUPERCAP. The average costs along with costs of supercapacitors found in the literature are depicted in Fig. 9. The total production costs for the AUTOSUPERCAP cells are in the range of 1.30–12.90 €/kW. These values depict the minimum and maximum costs of four types of researched cells, two AC based and 2 AC+MWNT based. Labour costs have been considered at European level. Material costs seem to be a considerable proportion of the total cost and, hence, have been further analyzed in Fig. 10. Figure 10 illustrates that the cost of the organic electrolyte $TEABF_4$ is the highest cost, followed by the cost of separator, the cost of cell packaging, and finally the cost of electrodes.

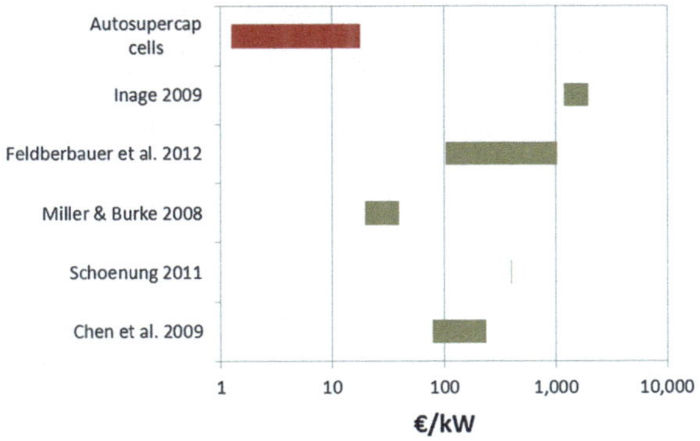

Fig. 9 Range of total production costs of supercapacitors taken from literature [7–11] and range of calculated total production costs of the Autosupercap pouch cells (includes the min. and max. prices for 4 types of researched pouch cells including AC based cells and AC+MWNT based cells) [12]

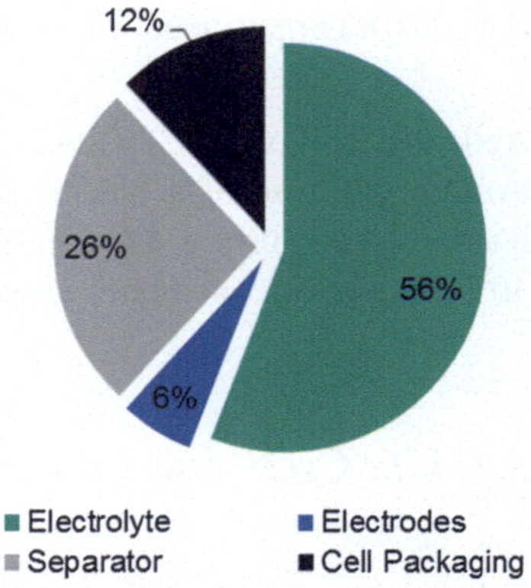

Fig. 10 Analysis of material costs for a typical supercapacitor cell of this study

4 Conclusions

The study proved that it is possible to fabricate small supercapacitor cells that can reach very high energy density over 40 Wh/kg while maintaining a reasonable power density around 6–10 kW/kg: these exceptional energy densities have been achieved by a combination of highly activated carbons, manufactured by MAST Carbon, and novel details in the material processing and cell fabrication techniques at the University of Surrey. Furthermore, addition of long, thin multiwall carbon nanotubes produced by Bayer Technology Services (BTS) raises the power density to 40 kW/kg while a very good maximum tested energy density is achieved of 27 Wh/kg. These levels of energy and power density are well above the levels of current commercial supercapacitors. A preliminary cost analysis for a supercapacitor cell based on the most promising technologies revealed cost levels in the range of 2–3.6 €/kW, which is within the limits set by the automotive industries. In this, materials cost featured at about 60 % of the total costs whereas the rest was mostly labour. The next phase of AUTOSUPERCAP involves scaling up of the supercapacitor cells where it remains to be seen whether this high performance and low cost will be maintained.

References

1. Lei C, Wilson P, Lekakou C (2011) Effect of poly(3,4-ethylenedioxythiophene) (PEDOT) in carbon-based composite electrodes for electrochemical supercapacitors. J Power Sources 196:7823–7827
2. Vermisoglou E, Todorova N, Pilatos G, Romanos G, Likodimos V, Boukos N, Lei C, Markoulidis F, Lekakou C, Trapalis C (2012) Few layer graphenes decorated with ag nanoparticles for supercapacitor applications. In: Proceedings of ECCM15, Venice, June 2012
3. Vermisoglou EC, Petridis D, Pilatos G, Romanos G, Likodimos V, Lekakou C, Trapalis C (2012) Iron carbide-graphene hybrid nanostructures. GrapHEL, Mykonos, pp 27–30

4. Todorova N, Vermisoglou E, Giannakopoulou T, Giannouri M, Lei C, Markoulidis F, Lekakou C, Trapalis C (2012) Simultaneous photoreduction and silver decoration of graphitic materials. GrapHEL, Mykonos, 27–30 Sept 2012
5. Lei C, Amini N, Markoulidis F, Wilson P, Tennison S, Lekakou C (2013) Activated carbon from phenolic resin with controlled mesoporosity for an electric double-layer capacitor (EDLC). J Mater Chem A. doi:10.1039/c3ta01638b
6. Weil M, Dura H, Shimon B, Baumann M, Zimmermann B, Ziemann S, Lei C, Markoulidis F, Lekakou T, Decker M (2012) Ecological assessment of nano-enabled supercapacitors for automotive applications. IOP Conf Ser: Mater Sci Eng 40:012013-1-6
7. Chen H, Cong TN, Yang W, Tan C, Li Y, Ding Y (2009) Progress in electrical energy storage system: a critical review. Prog Nat Sci 19:291–312
8. Schoenung SM (2011) Energy storage systems cost update. Sandia National Laboratories, Albuquerque
9. Miller JR, Burke AF (2008) Electrochemical capacitors: challenges and opportunities for real-world applications. Electrochem Soc Interface 17:53–57
10. Felberbauer K-P, Kloess M, Jungmeier G, Haas R, Könighofer K, Prüggler W, Pucker J, Rezania R, Beermann M, Wenzel A (2012) Energiespeicher der Zukunft, Energiespeicher für erneurbare Energie als Schlüssel-Technologie für zukünfitge Energiesysteme, Joanneum Research, Graz, Austria, Endbericht, 2012
11. Inage S (2009) Prospects for large-scale energy storage in decarbonised power grids. In: International Energy Agency, IEA, 2009
12. Dura H, Weil M (2013) Cost analysis of supercapacitor cell production. Proceedings of 2013 International Conference on Clean Electrical Power (ICCEP), Alghero, IEEE, pp 516–523

GREENLION Project: Advanced Manufacturing Processes for Low Cost Greener Li-Ion Batteries

Iratxe de Meatza, Oscar Miguel, Iosu Cendoya, Guk-Tae Kim, Nicholas Löffler, Nina Laszczynski, Stefano Passerini, Peter M. Schweizer, Franca Castiglione, Andrea Mele, Giovanni Battista Appetecchi, Margherita Moreno, Michael Brandon, Tadhg Kennedy, Emma Mullane, Kevin M. Ryan, Igor Cantero and Maxime Olive

Abstract GREENLION is a Large Scale Collaborative Project within the FP7 (GC.NMP.2011-1) leading to the manufacturing of greener and cheaper Li-Ion batteries for electric vehicle applications via the use of water soluble, fluorine-free, high thermally stable binders, which would eliminate the use of VOCs and reduce the cell assembly cost. The project has 6 key objectives: (i) development of new

I. de Meatza (✉) · O. Miguel · I. Cendoya
IK4-CIDETEC, Pº Miramon 196, 20009 Donostia-San Sebastián, Spain
e-mail: imeatza@grupocegasa.com

O. Miguel
e-mail: omiguel@cidetec.es

G.-T. Kim · N. Löffler · N. Laszczynski · S. Passerini
Helmholtz Institute Ulm, Karlsruhe Institute of Technology, Helmholtzstraße 11, 89081 Ulm, Germany
e-mail: stefano.passerini@kit.edu

P.M. Schweizer
Polytype Converting AG, 26, route de la Glâne, P.O. Box 1184, CH-1701 Fribourg, Switzerland
e-mail: Peter.Schweizer@polytype.com

F. Castiglione · A. Mele
Department of Chemistry, Materials and Chemical Engineering, Politecnico di Milano, Piazza Leonardo da Vinci, 32, 20133 Milan, Italy
e-mail: andrea.mele@polimi.it

G.B. Appetecchi · M. Moreno
ENEA (Italian National Agency for New Technologies Energy and Sustainable Economic Development), Technical Unit UTRINN-IFC, Casaccia Research Center, Via Anguillarese 301, 00123 Rome, Italy
e-mail: gianni.appetecchi@enea.it

E. Briec and B. Müller (eds.), *Electric Vehicle Batteries: Moving from Research towards Innovation*, Lecture Notes in Mobility,
DOI 10.1007/978-3-319-12706-4_4

active and inactive battery materials viable for water processes (green chemistry); (ii) development of innovative processes (coating from aqueous slurries) capable of reducing electrode production cost and avoid environmental pollution; (iii) development of new assembly procedures (including laser cutting and high temperature pre-treatment) capable of substantially reduce the time and the cost of cell fabrication; (iv) lighter battery modules with easier disassembly through eco-designed bonding techniques; (v) waste reduction, which, by making use of the water solubility of the binder, allows the extensive recovery of the active and inactive battery materials; and (vi) development of automated process and construction of fully integrated battery module for electric vehicle applications with optimized electrodes, cells, and other ancillaries. Achievements during the first 18 months of the project, especially on materials development and water-based electrode fabrication are reported herein.

Keywords Electric vehicles · Energy storage · Batteries · Alloys anodes · Water-based binders · Innovative processing · Battery manufacturing · Automation

1 Introduction and State of the Art

Society's current individual mobility behavior is creating a plethora of looming problems, such as fossil carbon intensity and the concomitant consequences regarding fossil resource supply or the emissions of pollutants such as nitrogen and sulfur oxides (NO_x, SO_2) and particulate matter. While pollutant problems can be addressed by catalytic converters and filters, expectations run high that the greenhouse gas and resource problems can be addressed by substituting internal combustion engine (ICE) cars with battery powered electric cars (BEV). Most of the major car manufacturers have announced BEVs as part of their product lines in the immediate future.

M. Brandon · T. Kennedy · E. Mullane · K.M. Ryan
Department of Chemical and Environmental Sciences, Materials and Surface Science Institute, University of Limerick, Limerick, Ireland
e-mail: Kevin.M.Ryan@ul.ie

I. de Meatza · I. Cantero
CEGASA Internacional, Artapadura 11, 01013 Vitoria-Gasteiz, Spain

M. Olive
RESCOLL, 8 Allée Geoffroy Saint Hilaire, CS 30021, 33615 Pessac Cedex, France
e-mail: maxime.olive@rescoll.fr

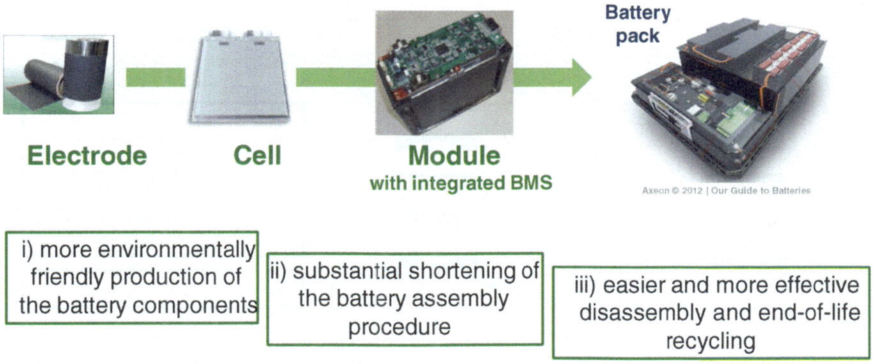

Fig. 1 Key levels of battery manufacturing value chain under development in GREENLION

Lithium ion batteries already dominate the consumer portable electronic and telecommunications market due to their higher power and energy density and they are also indicated as the option for the next generation of hybrid and electric vehicles (HEV, EV). The wide deployment of lithium ion batteries in the automotive industry would have tremendous consequences on the battery-market and it would further strengthen the central role of these systems in the field of energy storage. For that, considerable efforts are now focused on the development and realization of lithium ion batteries able to fulfill the requirement necessary for the application in HEV and EV. When the present lithium ion technology is considered, the safety and cost of batteries appear as the main drawbacks holding the introduction of this technology into the automotive market.

In order to tackle these issues from the manufacturing perspective before the final battery pack integration, the GREENLION consortium has identified three key levels in the value chain (Fig. 1), namely battery components, especially electrode processing, individual cells and battery modules, oriented to the battery pack for EVs.

As stated previously, the automotive industry is demanding for safe and low cost lithium ion (Li-ion) batteries to bring to the market higher range and affordable BEVs to substitute the ICE vehicles that won the battle a 100 years ago, due to Ford's mass production line, low oil cost and insufficient battery technology of the time.

The current Li-ion battery manufacturing process has advanced greatly in the last 20 years thanks to the consumer portable electronic industry and those developments are also the basis for the production of large format cells demanded for automotive application. The Li-ion battery production comprises a sequence of steps that can be summarized as electrode preparation, cell fabrication and battery module assembly, as presented in Figs. 1 and 2.

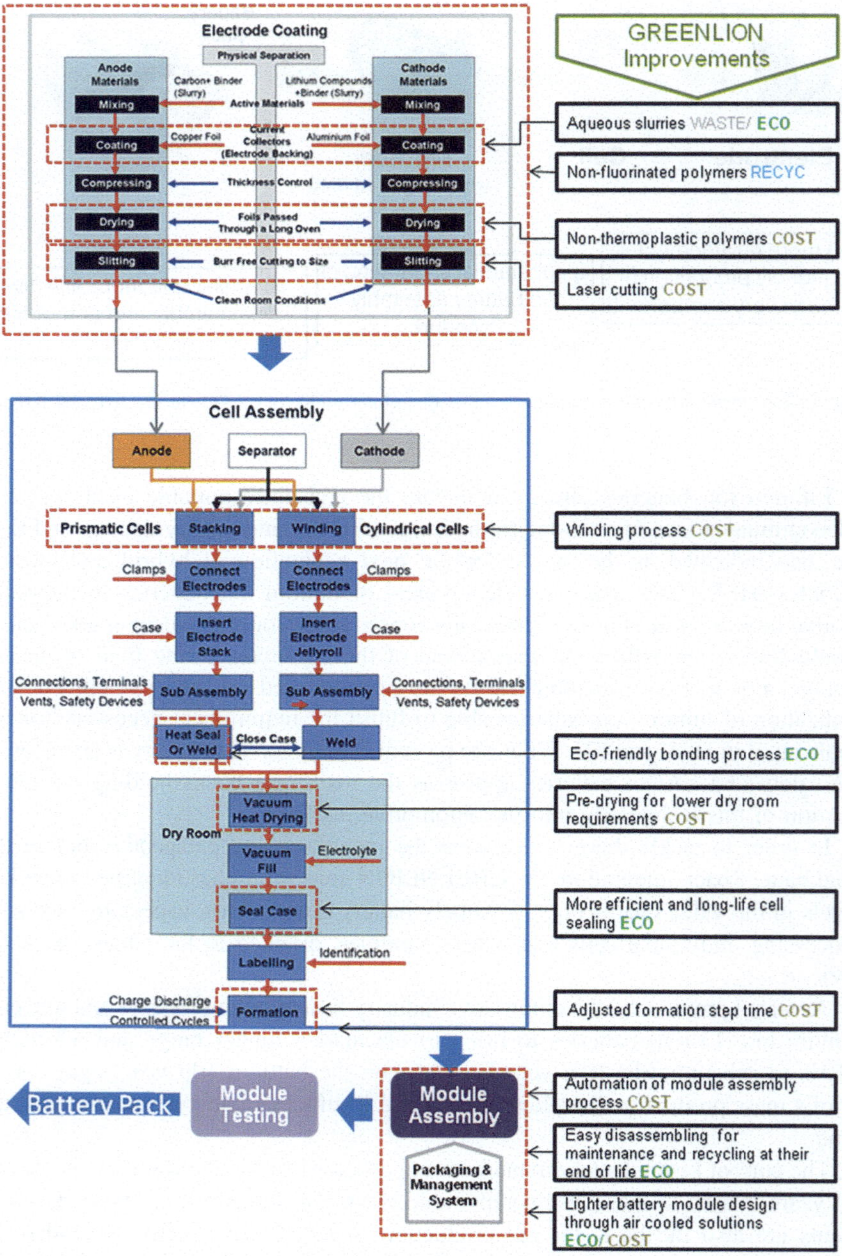

ECO: Eco-friendly; **COST**: Lower cost; WASTE: waste reduction; RECYC: easier recycling

Fig. 2 Breakdown of steps for Li-ion battery manufacturing and improvements proposed in GREENLION

1.1 Electrodes

The initial large scale automotive prototype lithium ion batteries use hard carbon or graphite as negative electrode materials, and nickel-substituted cobalt and/or manganese oxide (NCA, NMC), manganese spinel oxide or lithium iron phosphate as cathode active materials. Since for this first stage, the basic formulations and materials from consumer batteries are being used, the costs of the electrodes still need to be decreased for the final use in automotive batteries.

Numerous groups have been studying and developing new electrode and electrolyte materials with suitable characteristics and improved performance for the realization of greener and lower cost batteries, and promising results have already been achieved [1, 2]. However, to realize batteries with such kinds of properties, not only active and electrolyte materials have to be taken into account but, in general, all the battery components and even the process to realize the batteries need to be considered and improved. Apart from the continuous research on materials with increasing energy and power density, safety, durability and cycle life (targets of at least 30 % improvement for 2020), current estimations point to a 10–15 % reduction cost from the active materials that can reach up to 35 % through electrode process optimization [3].

Several studies are now also focused on the improvement of the inactive materials as well as of the electrode production. In this context, a key role is certainly played by the binder. As a matter of fact, this component is not only responsible for the binding of the active materials and the conductive agent to the metal current collectors, but it also strongly affects the electrode processing. Consequently, the improvement of the binder must necessarily be considered as a key point for the development of new safe and greener batteries.

An interesting example of the influence of the binder is observed when the preparation of electrodes based on lithium iron phosphate ($LiFePO_4$) is considered. $LiFePO_4$ displays high stability of the capacity during prolonged cycling; it is environment-friendly, cheap, and safe [4, 5]. Because of these characteristics, it is considered as a very attractive cathodic material. So far, however, most of the research and the development in composite cathodes have been focused on the use of fluorinated binders and practically all commercial lithium-ion batteries are made using poly(vinylidene fluoride) (PVdF) as the binder. However, this polymer is expensive (industrial cost in the multiton scale is around 15–18 EUR/kg); it requires the use of volatile organic compounds that are often toxic (like N-methyl pyrrolidone, NMP) in the processing, and it is not easily disposable at the end of the battery life. The introduction of alternative binders, as well as an improved preparation procedure, is necessary.

Recently, alternative binders have been introduced for the manufacture of anodes for lithium-ion batteries, like styrene butadiene rubber (SBR) that can be processed in water. Among them, one of the most interesting is certainly the sodium salt of carboxymethyl cellulose (CMC), which is a water-soluble material. This is certainly the greatest advantage of CMC because it allows processing in aqueous slurries rather than in polluting, health and environment unfriendly, volatile organic-compound-based slurries. The second great advantage of CMC resides in its easy disposability at the end of the life of the battery. Once the electrode is extracted, the active electrode material can be easily recovered by pyrolysis of the binder. Last but not least is the material cost. The CMC industrial price is about 1–2 EUR/kg, i.e., about 1 order of magnitude lower than PVdF.

1.2 Cell Assembly

Battery pricing is significantly impacted by material costs and manufacturing cost in mass production due to the multitude of operations and the precision required. The electrode thickness produced at the first step can range from 0.05 to 0.2 mm depending on the electrode type (cathode or anode), the intended application of the battery (high capacity or high power) and the cell design (cylindrical or planar).

Cylindrical cells, where components are staked and wound to be inserted into cylindrical cases, and prismatic cells, with stacked electrodes and separators, are currently the predominant designs. Pouch cells are prismatic cells with aluminum-polymer soft pack instead of metal can, so they achieve a packaging efficiency of 90–95 % and higher energy density. With high volume, any reasonable size can be economically produced. Lithium polymer pouch cells are increasingly being considered as alternatives to large prismatic cells for automotive applications; because their form is flexible they can be packaged more efficiently, and reduced cell packaging overheads result in high battery energy density. Due to large surface area and aspect ratio they have good heat dissipation. However, the cells have low mechanical stability and, therefore, more robust packaging is required.

1.3 Module Design and Assembly

In order to develop a battery module as a building block of a battery pack, first of all it is highly convenient to have as much information as possible about the characteristics of the vehicle to be powered by the energy stored in the battery in terms of weight, friction, aerodynamic coefficient, efficiency, voltage and current of the power train, … Besides, others features related to the vehicle performance must be defined, as the energy storage will be sized in order to cope with these requirements, such as autonomy, acceleration, maximum speed, cruise speed, etc., referred to a given driving cycle.

According to the vehicle characteristics and requirements, and once the cell has been selected, tested and modeled, all this information will be used to determine the required number of cells and modules and their series/parallel connection inside—respectively—the module and the battery pack, so that the required voltage, current, energy and power values are met. In order to define the optimum possible arrangement, the resulting module will be simulated—both electrical and thermally—out of the previously obtained models of the cells. Special consideration should be put in the lay-out of the cells since the thermal behavior of the module will strongly depend on this.

Most of the systems using batteries require a certain number of cells connected in series and parallel in order to achieve the desired voltages and current. Therefore, all the cells should be kept in the same state of charge (SOC) in such a way that the capacity of the resulting module or battery-pack is not reduced due to a weak cell that reaches the cut-off voltage sooner than the rest, or to an incomplete charge caused by a cell with a voltage higher than the others. Therefore, a cell balancing system that keeps the cells in the same SOC is required to improve the performance of the module.

Besides, special care must be taken in order to ensure that no cell is over charged or discharged, due to the electrochemical inequalities of the cells inherent to the manufacturing process or to uneven working or balancing conditions. Otherwise, the users' and cells' integrity could be compromised, as dangerous amounts of flammable gases and/or toxic chemicals can be released, and even end up in explosion.

In order to have an optimal use of the module, it is highly convenient to have access to the information concerning the state of charge (SOC, which is the remaining charge in the cells) and state of health (SOH, which is the capacity of the cells at a given time compared to that when they were new) of the cells. For all this reasons, a Battery Management System (BMS) is required in any system using lithium-ion cells.

The operation of batteries depends on an electrochemical process for both charging and discharging, and it is widely known that these chemical reactions are significantly dependent on temperature. Nominal battery performance is usually specified for working temperatures somewhere in the +20 to +30 °C range. However, the working temperature conditions of the cells can deviate substantially from nominal values, in such a way that batteries are operated at higher or lower temperatures. As a consequence, the performance of the cells is strongly affected: in general terms, discharge time (and therefore, capacity) decreases at lower temperatures, and the number of charge and discharge cycles is reduced when working at higher temperatures. Besides, and from a safety point of view, it is extremely important to avoid a thermal runaway (uncontrolled temperature increase) in the cells.

Therefore, a Thermal Management Systems (TMS) is required to maintain the cells within a safe temperature range that, besides, allows optimizing the performance of the module. In order to cool-down or heat-up the cells, different systems can be used, being air or liquid cooling the most usual choices.

2 Project Description

2.1 Project Approach and Objectives

In the GREENLION project (www.greenlionproject.eu), we address the issues cited previously by the industrial development of eco-designed processes at the electrode, cell and battery module level. At the **electrode processing stage** (that will be otherwise independent of the active materials chemistry), developing and making use of:

1. aqueous slurries rather than toxic organic volatile compounds (25 % cost reduction);
2. non-thermoplastic polymers that allow for high temperature drying, resulting in shorter and less expensive assembly procedure (10 % efficiency); and
3. easily disposable non-fluorinated polymers (at expected 10 times less materials cost).

At the **cell assembly level**, further improvements to the existing procedures as well as changes at some steps of the assembly process will be developed to increase energy efficiency and shorten times (and hence lower costs) during the manufacturing process, by implementing:

1. laser cutting instead of mechanical notching of the electrodes (15 % cost),
2. adjusted stack winding of components from aqueous-based electrodes and their drying process before electrolyte filling and sealing, to lower dry room requirements,
3. environmentally friendly bonding process for more effective and long-life cell sealing, and
4. adjusted formation step time (ideally for electrodes with reduced formation cycle) in cell manufacturing line (5 % time reduction).

Finally, developing a modular battery allows an easier handling of cells within a complete battery pack. At this **battery module level**, GREENLION project will design an autonomous unit including its own electrical and thermal management as a simple and reliable building block that will allow the manufacturing and maintenance of the whole battery packs easier and more inexpensively, with the lowest possible environmental impact. This will be achieved by:

1. lighter battery module designs (including electronics) with the possibility of implementing air cooled solutions instead of liquid cooling systems (expected 20 % less weight),
2. bonding process of module housing for safe operation but easy disassembling for maintenance and reuse/recycling at their end-of-life, and
3. automation of module assembly process (3 s/cell vs. manual assembly).

These developments will be scaled-up and realized in pilot lines during the project, following a continuous environmental assessment of materials and processes. A validation of the finally assembled battery module will be carried out by the automotive end-user who will also provide the targets and specifications for (H)EV application.

General project approach and objectives are summarized in Fig. 2. Progress beyond current State of the Art is also indicated.

2.2 Project Consortium

The scientific and technological cooperation in GREENLION consortium and their roles in the project are well balanced covering the complete chain from raw material, scientific comprehension, technological research and end users. As an essential part of the project, the industrial partners will commit to exploit all commercial aspects of the new manufacturing processes. To this purpose, the industrial partnership was designed to combine Li-ion cells and module manufacturers (CEGASA), processing equipment manufacturers (POLYTYPE and KEMET), material suppliers (SOLVAY and TIMCAL), automation of assembly processes providers (MONDRAGON ASSEMBLY), recycling and waste treatment services suppliers (TECNICAS REUNIDAS), and car manufacturers (SEAT and VOLKSWAGEN). The research institutions (CIDETEC, ENEA, RESCOLL, AIT) and universities (University of Muenster, Politecnico di Milano, University of Limerick) in GREENLION consortium provide complementary skills and expertise in the relevant fields of research and development that are necessary to achieve the project objectives.

3 Outcome of the Project

3.1 Baseline for GREENLION Project and Performance Indicators

Knowledge will be generated well beyond state of the art and the limitations of current Li-ion battery manufacturing process. In particular, Table 1 summarizes and quantifies the most significant targets.

3.2 Expected Impact of the Project

GREENLION will provide advances to a number of scientific and engineering challenges for battery cell and module manufacturing, and their performance thereof. The successful resolution of these will lead to breakthroughs in automotive lithium ion batteries for electric vehicles and thus to the development of a sustainable mobility and quality of life.

Table 1 Summary of most relevant GREENLION performance indicators and targets

	Proposed innovation-performance indicators
Electrode	Development of innovative electrodes realized by water-based processes to realize electrodes characterized by: (i) high thermal stability to allow high-temperature drying (>150 °C) in order to allow assembly in less stringent dry room operating conditions and reduce post-coating treatment time); (ii) high electrochemical stability to allow the use of high voltage cathodes (at least 5 V vs. Li/Li^+); (iii) high capacity retention upon cycling (more than 80 % of initial capacity after 1,000 cycles)
	The final goal is to obtain anodes and cathodes for lithium-ion batteries with storage capacities as high as, respectively, 300 and 150 mAh/g (excluding the weight of the current collectors), and surface loadings of, at least, 5 mAh/cm^2
Cell	Implement laser cutting/slitting instead of mechanical notching of the electrodes achieving negligible degradation of active material in the cut area, reduction of burrs resulting in a safer cell, expected 15 % cost saving due to reduced maintenance and higher process efficiency
	Based on such innovative electrodes, GREENLION proposes to design and develop cells capable of delivering a specific energy of 200 Wh/kg, which is the actual target for automotive applications
Module	Lighter battery module designs (including electronics) by evaluating the implementation of air cooled solutions instead of liquid cooling systems (20 % less weight)
	Automation of module assembly process with a handling time down to 3 s per cell) will enable cost reduction and quality in line with what achieved in the highly automated cell manufacturing

Greening our transport system is necessary not only to avoid the influence of oil supply ($147 per barrel peak in 2008) but also to achieve EU and international targets in emissions reductions. In the EU, 19 % of total greenhouse gas emissions and 28 % of CO_2 emissions in 2005 are linked to the transport sector. More than 90 % of the total EU transport-related emissions are due to road transport. While total EU emissions declined, transport emissions increased continuously between 1990 and 2005 due to high growth in both passenger (28 %) and freight transport (62 %).

Current and near-term (i.e. Li-ion) battery technology development is one of the key factors on the Mobility Electrification and the large scale production of these automotive batteries and reducing their costs is, in fact, critical for market entry and acceptance of Electric Vehicles. In order to achieve a break-even cost with internal combustion engines, battery costs must be reduced from the current estimated range of 675–500 € per kilowatt-hour (kWh) at high volume production (order of 100 k units) down to 350–275 €/kWh by 2020. R&D to improve power (W/kg) and energy density (Wh/kg) in order to increase driving autonomy, reductions in recharge time and achieving life cycles that approach vehicle life spans is also imperative. Increasing production rate from 10,000 to 100,000 batteries/year reduces cost by ~30–40 % [6].

GREENLION addresses further reduction costs driven not only by high volume manufacturing, but also from the components processing conditions. The use of water based binders, an order of magnitude cheaper than conventional fluorinated ones will drive down the cell manufacturing costs, besides being more environmentally friendly and eco-sustainable at the end of life of the cells. Besides the improvement in environmental, health and safety terms (including "working-condition-friendly" considerations), the initial inversion and running costs of the solvent recovery system would be avoided and water is indeed cheaper than NMP. Even though in current production plants the recovered NMP is purified and offered again at 50 % of the cost of pure solvent, distilled water is also cheaper (0.20 €/L) than 50 % of recycled NMP 0.90 €/L (~ 1.8 €/L pure).

Expected impact in the field of new competitive processes, by means of production automation is also foreseen. Not only will the results of the project efforts enable lower cost and greener lithium battery packs production, but also equipment manufacturing and high added value processes will be developed. These new automated processes will contribute to a substantial cost reduction of lithium battery packs, and will facilitate their introduction to mass production.

Automation and new process development will improve the quality and yield of the production, while at same time reduces labor costs per kWh. This project, with the development of the specific equipments for module assembling, will enable a cycle time of 3 s for each cell. This results in a module production capacity of 880 MWh/year.

Globally, automation and equipment development in this project will enable a cost reduction of the whole battery pack of 15 %. Having in mind that only 24 % of the cost is related to the module/pack manufacturing (60 % are materials components and 16 % are transports and others), it represents a major step in the way to mass production. Market growing will also pull down the prices of the materials, and it will open the way to the mass production at competitive costs.

3.3 Results Achieved

GREENLION is currently at month 18 of a 4 year-long workplan. During this first stage of the project, efforts have been mainly focused on the electrode processing step, with the development and testing of active materials and binders suitable for water-based slurry formulations and electrode coating process. First selected formulations have been used for small scale GEN1 prototype pouch cell assembly while the optimized module design and assembly process is underway. These results and advances are summarized in the following sections.

3.3.1 Materials Development and Water-Based Electrode Processing

Among the main research topics of the GREENLION Project are to be highlighted the development of ionic liquid-based electrolytes and the realization of electrodes, prepared through innovative, eco-friendly process routes, based on high-voltage cathode and large-capacity anode materials. There is growing up interest in replacing the organic solvents currently used in lithium batteries [7–9] with ionic liquids, ILs, since their non-flammability and negligible vapor pressure in conjunction with wide chemical, electrochemical and thermal stability, high ionic conductivity and heat capacity. Our basic idea is to favorably combine different IL sets in order to obtain ionic liquid mixtures with improved performance. For instance, N-methyl-N-propylpyrrolidinium bis(fluorosulfonyl)imide ($PYR_{13}FSI$) was found to exhibit moderate viscosity and low melting point, allowing fast ion conduction even at low temperatures [10]. On other hand, the much cheaper N-methyl-N-propylpyrrolidinium bis(trifluoromethanesulfonyl)imide ($PYR_{13}TFSI$) shows wider thermal and electrochemical stability [11]. In order to verify if these characteristics could be combined, $PYR_{13}FSI$-$PYR_{13}TFSI$ mixtures were prepared and investigated in terms of NMR spectroscopy, transport properties and density measurements.

Remarkable conduction values, e.g., approaching 10^{-3} Scm^{-1}, are achieved already at −20 °C for mole fraction ranging from $0.6 \leq x \leq 0.8$ whereas both the raw ionic liquid materials ($PYR_{13}FSI$ and $PYR_{13}TFSI$) are solid at this temperature (see Fig. 3). This highlights the synergic effect exhibited in ionic liquid mixtures, especially for low temperature applications.

NMR heteronuclear NOE correlation experiments (HOESY) experiments have been successfully used for the assessment of the intermolecular contacts among the F atoms of the anions and the H atoms of the cations in pyrrolidinium based ionic liquids, thus providing information on the local structural organization. The experiments have shown a peculiar cation-anion organization in the three investigated blends responsible of their favorable physico-chemical characteristics.

High nominal voltage cathode materials, combined with large capacity anodes are appealing issues for the realization of lithium batteries with high gravimetrical and volumetric energy. In this first stage of GREENLION, cathodes based on $LiNi_xMn_yCo_{1-x-y}O_2$ (NMC) and anodes based on carbonaceous materials (graphite, SLP) have been developed. The composite electrodes were fabricated using the fluorine-free, water-soluble, natural binder carboxymethylcellulose sodium salt (CMC) instead of the more expensive and less environmentally friendly polyvinyliden-di-fluoride (PVdF) in N-methyl-pyrrolidone (NMP). The use of CMC allows also easier recycling of the battery components. For instance, the dissolution in water of the binder allows, for example, a full recovery of the metallic current collectors [12, 13].

Commercially available NMC cathode and Timcal SLP 30® graphite anode tapes based on the aqueous CMC binders were prepared using a pre-pilot automated coating line. The cycling performance tests (Fig. 4) evidenced a time-stable capacity of 130 mA h g^{-1} for more than 40 cycles with coulombic efficiency higher than 99.0 % for the NMC cathodes. The SLP 30® anodes showed very good performance in terms of reversibility of the intercalation process. The specific capacity leveled 375 mA h g^{-1} after a few cycles. Upon 80 cycles, the SLP 30® electrodes showed still high cycling stability and columbic efficiency above 99.9 %. These results support for a further development of the aqueous CMC binder-based electrodes.

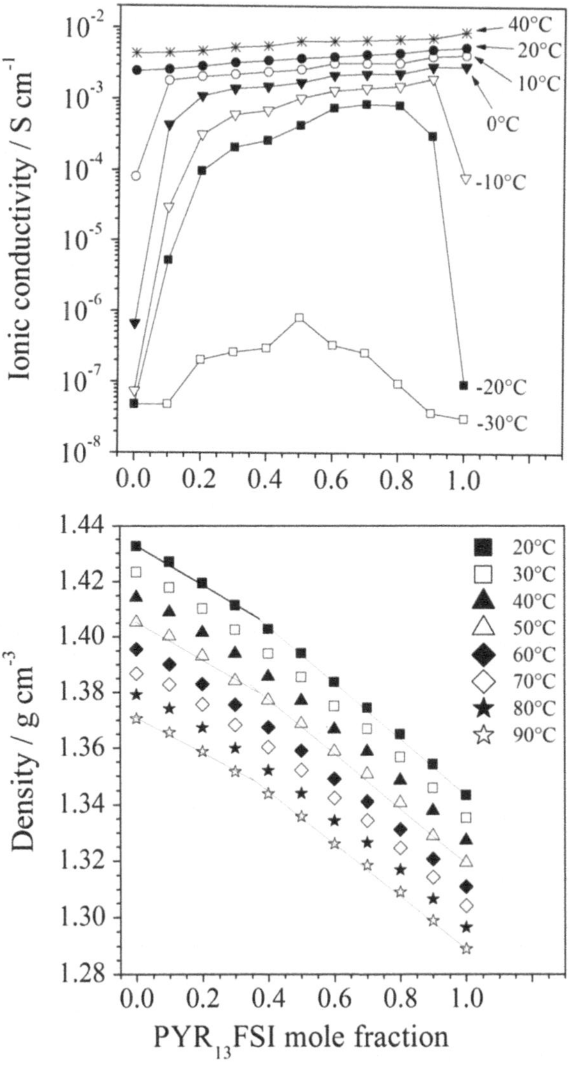

Fig. 3 Ionic conductivity (*upper panel*) and density (*lower panel*) versus $PYR_{13}FSI$ mole ratio dependence for (x)$PYR_{13}FSI$/(1−x)$PYR_{13}TFSI$ binary electrolyte mixtures at different temperatures

In addition, alternative water-soluble binders have been studied. Impressive electrochemical performance has recently been reported for Si nanopowder [14] and nanowire [15] anodes prepared from aqueous slurries using 15 wt% alginate as binder. To date, however, no study has been undertaken using alginates as the binder for graphitic anodes, which are almost ubiquitous in present lithium ion battery technology. It is apparent from Fig. 5 that the graphite anode with 7.5 wt% alginate outperforms that with 10 wt% PVDF (a common commercial level) over the course of the first 65 charge/discharge cycles. This result suggests that alginate may be a suitable candidate for aqueous manufacturing of anodes.

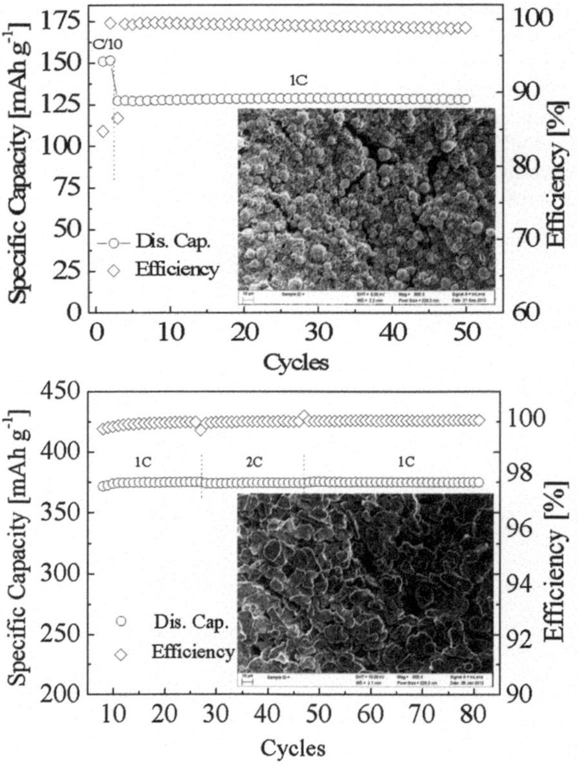

Fig. 4 Cycling performance versus Lithium and SEM image (*inset*) of NMC cathode (*upper panel*) and SLP 30® anode (*lower panel*) in (1 M) $LiPF_6$/EC:DMC (1:1 in wt/wt) electrolyte at 20 °C. Current rate: 0.1–1 C; mass loading: NMC, 3.29 mg cm^{-2}; SLP 30®, 2.45 mg cm^{-2}

The CMC based formulations will be the first to be trialed in pilot line in order to develop optimized coating machinery and electrodes for cell assembly. The most efficient way of manufacturing battery electrodes is to simultaneously coat both sides of the substrate and to use a flotation dryer for removing the solvent. This configuration requires one of the coatings to be applied in the so called kiss coating mode as depicted in Fig. 6 for the slot coating process.

Coating trials have been carried out on a pilot machine, allowing the adjustment of parameters to achieve an excellent uniformity of the kiss-coated layer, i.e. by suppressing cross lines generated by web flutter in the flotation dryer, and by suppressing longitudinal bands generated by web deformations upstream of the slot die.

3.3.2 Cell Assembly and Module Design

During the first year, GEN0 prototype cells (10–14 Ah) were assembled as baseline for the project, from electrodes prepared with commercially available water-soluble binders and graphite/$LiFePO_4$ (C/LFP) chemistry.

The NMC and SLP 30® electrodes (around 1 m^2) prepared in a pre-pilot automated coating line were used to assemble GEN1 small pouch cells (0.5–1.5 Ah) as shown in Fig. 7, following the first large cell design (30 Ah target)

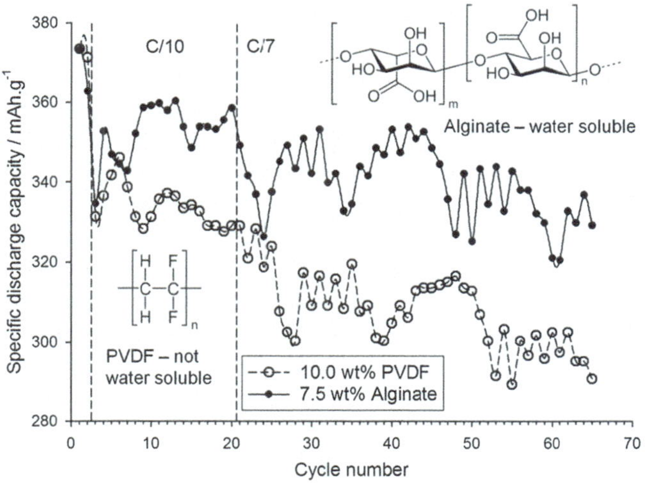

Fig. 5 Comparison of the specific discharge (delithiation) capacities of anodes prepared using polyvinylidene fluoride (PVDF) or alginate binders. The only other component of the anodes was the active graphite material—TIMREX® SLP30 by TIMCAL. The first and second charge/discharge cycles were conducted at slow rates of C/40 and C/25 respectively to facilitate the formation of a stable SEI layer. Voltage limits were between 5 mV and 1.5 V versus Li/Li^+. The electrolyte was 1 M (EC:DMC, 1:1 v/v) and the experiments were conducted at ambient room temperature

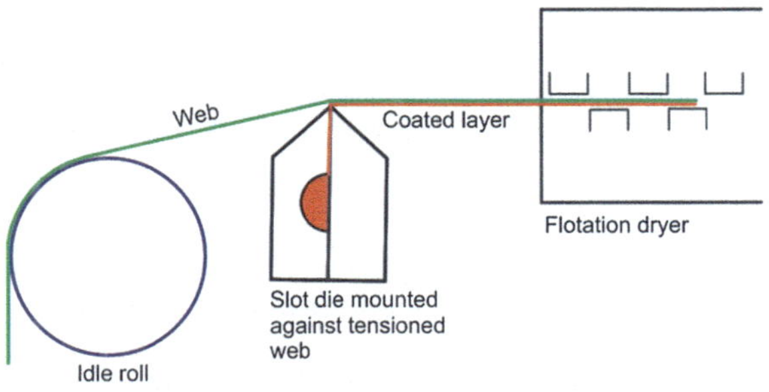

Fig. 6 Schematic view of coating the web-underside by the slot coating process operating in the kiss or tensioned-web mode

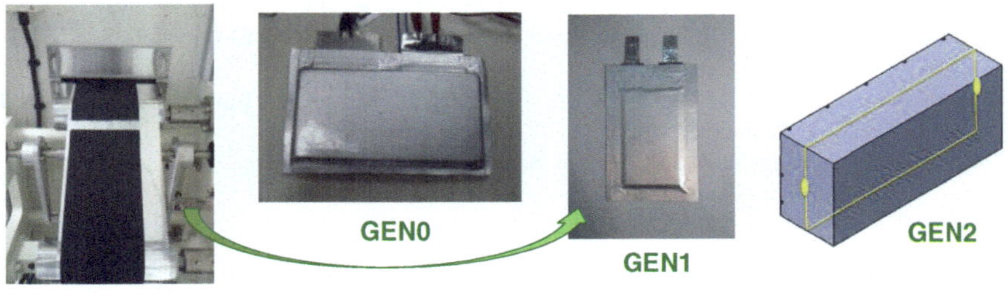

Fig. 7 From *left* to *right* CMC-based electrodes prepared in pre-pilot coating line, *GEN0* C/LFP cell, *GEN1* small pouch cell with SLP 30® and NMC electrodes and schematic *GEN2* power cell and module design

proposed to fulfill the energy requirements of the end-users for an efficient automotive battery module.

While laser notching trials of electrodes with both PVDF and water-based binders are underway, conventional cutting dies (mechanical notching) and manual stacking process were used for GEN0 and GEN1 cell assembly. Automated stacking-winding will be implemented for the GEN2 cell that has been adopted as the most efficient electrical and thermal design for high power performance.

The design of a lighter battery module suitable for automated assembly and easier disassembly is ongoing, coupled to the GEN2 power oriented cell design. Different aspects such as minimum mechanical fitting by the assembly process, modular assembly including liquid cooled cold plates, mechanical absorption of cell swelling and venting are under consideration.

References

1. van Schalkwijk WA, Scrosati B (2002) Advances in lithium-ion batteries. Kluwer Academic, New York
2. Nazri G-A, Pistoia G (2004) Lithium batteries. Kluwer Academic, New York
3. Joint European Commission/EPoSS/ERTRAC Expert Workshop (2009) Batteries and storage systems for the fully electric vehicle, 19 June 2009
4. Arnold G, Garche J, Hemmer R, Ströbele S, Vogler C, Wohlfahrt-Mehrens M (2003) J Power Sources 247:119–121
5. Striebel K, Shim J, Sierra A, Yang H, Song XY, Kostecki R, McCarthy M (2005) J Power Sources 146:33
6. Howell D (2010) DOE annual merit review meeting
7. Shin J-H, Henderson WA, Appetecchi GB, Alessandrini F, Passerini S (2005) Electrochim Acta 50:3859
8. Appetecchi GB, Montanino M, Balducci A, Lux SF, Winter M, Passerini S (2009) J Power Sources 192:599
9. Balducci A, Jeong SS, Kim GT, Passerini S, Winter M, Schmuck M, Appetecchi GB, Marcilla R, Mecerreyes D, Barsukov I, Khomenko V, Cantero I, De Meatza I, Holzapfel M, Tran N (2011) J Power Sources 196:9719
10. Appetecchi GB, Montanino M, Balducci A, Lux SF, Winter M, Passerini S (2009) J Power Sources 192:599
11. Appetecchi GB, Montanino M, Carewska M, Moreno M, Alessandrini F, Passerini S (2011) Electrochim Acta 56:1300
12. Kim GT, Jeong SS, Joost M, Rocca E, Winter M, Passerini S, Balducci A (2010) J Power Sources 195:6130
13. Lux SF, Schappacher F, Balducci A, Passerini S, Winter M (2010) J Electrochem Soc 157(3): A320
14. Kovalenko I, Zdyrko B, Magasinski A, Hertzberg B, Milicev Z, Burtovyy R, Luzinov I, Yushin G (2011) Science 334:75
15. Ge M, Rong J, Fang X, Zhou C (2012) Nano Lett 12:2318

Lightweight and Integrated Plastic Solutions for Power Battery Racks in Electric Vehicles

Thierry Coosemans, Javier Sanfélix, Maarten Messagie, Joeri Van Mierlo, Anthony Alves and Gilles Waymel

Abstract The Operating Energy Racks for Full Electric Vehicles project (OPERA4FEV) is a European project under the 7th Framework Program of the European Commission. The project started in September 2011 for a total period of 54 months and aims to propose a cheap, light and versatile alternative solution to the present metal-based technology for power battery racks in electric vehicles. It also aims for a high level of function integration while taking crash and safety regulations into account in the mean time. It involves a consortium of 10 partners from 6 European countries and has a total budget of €7 millions.

Keywords Advanced battery racks · Life cycle assessment · Eco-design · Mass production

T. Coosemans (✉) · J. Sanfélix · M. Messagie · J. Van Mierlo
MOBI - Mobility, Logistics and Automotive Technology Research Centre, Vrije Universiteit Brussel, Pleinlaan 2, 1050 Brussels, Belgium
e-mail: Thierry.coosemans@vub.ac.be

J. Sanfélix
e-mail: jsanfeli@vub.ac.be

M. Messagie
e-mail: maarten.messagie@vub.ac.be

J. Van Mierlo
e-mail: joeri.van.mierlo@vub.ac.be

A. Alves · G. Waymel
Mecaplast, Group Centre Technique de Lens, PA La Croisette, Rue des Poissonniers, 62300 Lens, France
e-mail: aalves@mecaplast.com

G. Waymel
e-mail: GWaymel@mecaplast.com

E. Briec and B. Müller (eds.), *Electric Vehicle Batteries: Moving from Research towards Innovation*, Lecture Notes in Mobility,
DOI 10.1007/978-3-319-12706-4_5

1 Introduction

Growing awareness of the upcoming oil shortage and global warming urges the automotive industry to develop new and environmentally friendly solutions. Currently, electric vehicles form the most promising technology for a more sustainable road transport [1]. Global accords, like the Kyoto agreements on the emission targets in 2020, lead to more and more restrictions for classical ICE-based vehicles. In October 2009 the European Council set the appropriate abatement objective for Europe at 80–95 % below 1990 levels by 2050 [2]. This roadmap demonstrated that there is a need of 95 % decarbonisation of the road transport sector as well as of the power sector. After 2020, further (more than 30 %) engine efficiency improvements are limited and relatively costly, while the amount of biofuels that will be available may be limited [3]. Hence the electrification of passenger cars is required. Electric vehicles can be fuelled by a wide variety of primary energy sources—including gas, coal, oil, biomass, wind, solar and nuclear—reducing oil dependency and enhancing energy security. Social and demographic evolutions are favoring the use of electric vehicles as well. In all developed and developing countries the population becomes more and more urban: by 2030, 65 % of the global population will live in urban areas, and by 2050 the urban population will achieve 70 %.[1] The growing urban population will lead to the further development of urban policy measures (like in London or Paris) restricting the presence of highly pollutant vehicles in city centers, and favoring silent and clean vehicles by for instance offering free parking space, the use of reserved lanes for public transport or the exemption of toll and congestion taxes. In addition it will lead to new mobility systems in which travellers are changing from car owners to customers of mobility services comprising various sustainable means of transport such as trains, subways, tramways and shared electric vehicle fleets. In addition, it has to be noticed that currently daily average commute distance is 30 km at an average speed of 30 km/h what is easily within the reach battery electric vehicle's drive range. It is hence foreseen that the market share of electric vehicles (hybrids, plug-in hybrids and pure battery-electric vehicles) will steadily grow in the next decades.

Although electric vehicles have been evolving rapidly during the last years, improvements are still necessary in order to boost the introduction of these vehicles into the market. This is especially true for the battery rack. Not only is there a strong urge to develop battery cells with increased specific power and energy and an improved cycle life; highly integrated and cost-effective battery packs need to be further developed as well. Currently battery packs are still expensive and leave room for a clear improvement on environmental impact, suitability for mass-production, weight and safety. In OPERA4FEV the Life Cycle Assessment (LCA) methodology will have an important role; the results of the different LCA tasks will guide the project in the ecodesign of the rack, and therefore to develop an

[1] World Urbanization Prospects, United Nations.

environmentally friendly product. This project aims to develop such battery packs, integrating hydraulic connections and component housing in a thermoplastic approach to reduce cost, weight and assembly time.

2 The Project

The OPERA4FEV project aims to develop thermoplastic battery racks based on Li-ion technology for two functional demonstrators: one for a large scale vehicle and one for a "niche" car or city vehicle. To improve deployment of electrical vehicles in Europe, large scale production processes for racks and electrical components need to be developed. OPERA4FEV also aims at improving the manufacturing process of benchmark battery racks in a more eco-friendly way. In order to obtain a product with better integration characteristics as the current one and also with enhanced environmental competences, the currently used metal is being replaced by thermoplastic as a main component and all the parts involved in the rack (cooling system, printed circuit boards…) are being redesigned (Fig. 1).

In order to show the relevance of the solution and meet strong industrial benefits, the OPERA4FEV thermoplastic racks will be manufactured and assembled with industrial means. The OPERA4FEV project aims to achieve full thermoplastics battery racks that integrates electrical and hydraulic connections and adaptable to any full electric vehicle model. Manufacturing processes are considered from the production of the cells itself up to large-scale system. Special attention is paid to thermal management systems and safety issues, which are critically dependent on cells system design and cells properties. As safety of the on-board battery in electric, hybrid, and plug-in vehicles is of paramount importance to the international automakers, OPERA4FEV pays particular attention to evaluate the effects of the rack characteristics regarding vehicle crash safety, and focuses on the potential risks for the vehicle and its occupants in case of failure of one or more batteries.

Plastic material is there a key element. Several grades of plastic will be used in the rack depending of their core function. Highly reinforced plastics will be used for strength. High temperature plastics will be used for insulation. Flame retardant plastics will be used for fire resistance.

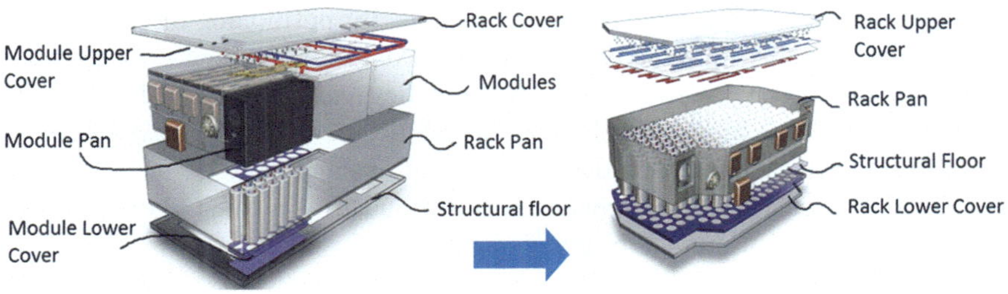

Fig. 1 Metal versus plastic battery rack assembly

Each plastic casing, support, bracket, separator or small part will answer to a single or several functions. Then the plastic grade will be selected regarding the dedicated aim of each part.

During the development process, the project takes into account the availability of critical materials and dismantling/recycling methodologies. This will lead to a complete eco-designed solution, able to be proposed at large scale. In addition, OPERA4FEV concept is fully in line with the European end of life vehicle directive stating that by 2015, vehicles must be constructed of 95 % recyclable materials, with 85 % recoverable through reuse or mechanical recycling and 10 % through energy recovery or thermal recycling, the use of "green" composites is expected to greatly increase and prevail in the future vehicle components' development. The impact of the newly developed rack technology on the environment is being assessed by means of a Life Cycle Assessment (LCA).

Finally, the whole value chain for power battery racks is being addressed, including the eco-design (dismantling and recycling of critical materials), assembly and integration of cells and electrical components. This is clearly reflected in the available skills, expertise and competences of the project consortium. Summarized, the main OPERA4FEV aims to achieve the following goals:

- Integration of battery cells into the thermoplastic rack
  Advantage of plastic design is the possibility to integrate plastic shapes for fixations and fittings directly in the main casing casting. Shapes like clips, clamps, holes for screws can be directly obtained from molding. This can be applied to the cells. Also can be integrated the cells separators, insulators or cooling deflectors. Fixation for other subcomponents such as channels for wire harness or hollow ducts for air channeling to cooling system, can also be integrated. For after-sale maintenance, dedicated shapes can also be forecasted in order to accept bolt and screw fixation in case of breakage of the original fixation shape.
- Easier mounting and fast connections of cells
  Unitary sub components such as electronic cards, power connectors to cells or wire harnesses can be pre- assembled together on sub-assemblies item. This way, the pre-assembly is faster and can be automated. Then, the fitting of those sub-assemblies is quick and straightforward.
- Reduction of assembly time
  The automation of sub-assembly manufacturing allows a gain of time. The plastic material allows some dedicated supports for positioning of components. Then the assembly time can be dramatically reduced. Regarding cells assembly in the rack, plastic material insulation properties and freedom of shape can be used to allow the fitting of cells with a high degree of safety, confidence and reliability, and therefore improve the speed of assembly.
- 25 % Cost reduction on components (cells excluded) compared to existing rack
- 50 % Reduction of number of components
  The way plastic material can be shaped will allow to integrate fixation devices that otherwise would have been bought and assembled separately. Also, ducting

and piping will be integrated as hollow shapes in the main rack plastic casings and not assembled separately.
- Improvement of dismantling for better maintenance
- Eco-design of the racks and easier end of life (materials, dismantling) based on life cycle assessment Plastic material, at the end of life, can be more easily cut, broken, dismantled and grinded. If necessary, some fuse areas can be forecasted in the plastic shape in order to make the dismantling easier still.
- Plastics parts design to improve thermal regulation, guaranty safety/crash behaviour and deal with large dimensions
  Plastic material will be used as insulation between cells. The low thermal conductivity and the thermal inertia properties here being an advantage. Then, plastic design allows for dedicated internal piping and shapes whether a liquid or air cooling system is required.
- Use of recycled polymers (70 % in weight)
- 30 % Weight reduction on components (cells excluded)
- Concept proposal adaptable to automotive industry and evaluated with a representative tools and assembly line (able to reach 20 vehicle/day).

3 The Consortium

The consortium built around the OPERA4FEV project is perfectly representative of the whole value chain for the manufacturing of advanced battery racks in the near future. In order to ensure industrial relevance and impact of the research efforts, OPERA4FEV includes active participation of industrial partners, including SMEs, components suppliers, electrical vehicles manufacturers and component recyclers, as well as research institutions.

- *MECAPLAST Group* (*France and Monaco*) is the coordinator. Automotive equipment for plastic parts. It's in charge of the design of the part and coordination of the project.
- *CENTRO RICERCHE FIAT* (*Italy*), development and transfer of innovative content and car maker. It's in charge of the vehicle supply, assembly in the vehicle and tests on road conditions.
- *CETHIL* (*France*) research laboratory in thermal science and energy engineering. Research unit of INSA-Lyon, of Claude Bernard University and of CNRS. It's in charge of the cooling dimensioning, cells characterization and thermal tests.
- *VRIJE UNIVERSITEIT BRUSSEL* (*Belgium*) premier electric vehicle research facility in Belgium. VUB carries out the life cycle assessment of the newly developed rack technology.It's in charge of LCA analyses and eco-design.
- *EVE System* (*France*) engineering company, expert in design, prototyping and industrialization of electric architectures for electric and hybrid vehicles. It's in charge of electrical management, electrical component design and assembly of the complete racks.

- *OLESA* (*Portugal*) is an engineering and tooling company. It's in charge of near-industrial tooling supply.
- *GRUPO REPOL* (*Spain*) providers and custom compounders of Engineering Plastics (PA, PP, PC, PC/ABS, PBT, ABS, POM and PPS) for injection molding and extrusion. It's in charge of the material supply.
- *UNIVERSIDAD POLITÉCNICA DE MADRID* (*Spain*) technical university. It's in charge of crash calculation and vehicle crash test.

4 LCA and Eco-Design

To carry out the redesign of the mentioned rack the LCA has an important role in the whole project. LCA is a tool that assists in identifying opportunities to improve the environmental performance of products at various points in their life cycle, and informing decision-makers in industry, government or non-government organizations on this matter [4]. The European Commission (EC) and others (member states, industry, consumer organizations and environmental organizations) recognize LCA as "the best framework for assessing the potential environmental impact of products" [5]. A clear example that corroborates the importance given to LCA by the EC is the creation of the "European Platform on Life Cycle Assessment" (EPLCA) in 2005. The EPLCA focuses on providing coherent and quality-assured life cycle data, methods, and information [6], and therefore promote scientific robustness in Life Cycle Thinking practices.

A sound and solid LCA methodology has a series of defined phases which have to be followed and are described in the ISO 14040 and 14044. They are explained in more detail in the ILCD Handbook. An LCA study is divided in: goal and scope definition, inventory analysis, impact assessment and interpretation.

- *Goal and Scope definition*: The goal and scope of the LCA is defined considering the stakeholders' profiles and requirements. A functional unit has to be defined; the functional unit is a central element in an LCA it names and quantifies the qualitative and quantitative aspects of the product or system being studied. The system boundaries have to be described in this phase of the LCA, these boundaries define what phases of the life cycle are going to be assessed in the study and they vary depending on the goal and scope.
- *Inventory analysis*: All the processes that are inside the limits defined in the previous phase are part of the inventory needed to run the LCA. It implies the gathering of the data regarding the material and energy flows involved in the different processes. The data collected has to represent the system studied in the most accurate way.There are two types of data that are used in LCA, primary and secondary. Primary data is obtained from the manufacturer of the product, the secondary data is provided by national databases, consultants or research

groups. The data gathered is introduced in software where the system studied is modeled; the data will be also treated and adapted to the functional unit.
- *Impact Assessment*: In this part of the LCA the inputs and outputs of elementary flows that have been collected and reported in the previous phase are translated into impact indicator results related to human health, natural environment and resource depletion.
- *Interpretation*: The results of the LCA are evaluated in order to answer questions posed in the goal definition. The interpretation relates to the intended applications of the study and it is used to develop recommendations.

An LCA study is an iterative approach where these four phases are repeated as the study progresses, the following figure outlines the iterative process among the different phases of an LCA study (Fig. 2).

The role of the LCA methodology in eco-design projects is of great relevance, and hence it is one of the main aspects in the Opera4FEV project. Three tasks involving this methodology are included in the project: Quick Ecoscan, Full LCA and a Recyclability study.

- *Quick Ecoscan*: At the early stages of a project where the aim is to manufacture and improve an already existing product (battery rack in Opera4FEV), it is important to assess the impact in the environment of the different design options from the very beginning, hence it is guaranteed that the project will follow patterns focused on the design of product with a lower environmental impact. According with the general aim of OPERA4FEV (product improvement) the ILCD Handbook recommends to perform an identification of Key Environmental Performance Indicators(KEPI) of the product (simplified LCA). This recommendation is followed for the applied Ecoscan study in the project. A KEPI study performed at the beginning of eco-design projects it is useful for designers. Such studies empower designers with indicators that account the physical and chemical characteristics of the product, like global warming potential (GWP) or energy demand among others. Normally the indicators used are easy to interpret and require little data and time for calculation.
- *Full LCA*: A complete LCA study is being carried out. Here the weight reduction applied in the battery rack will be assessed in all the life phases of a full electric vehicle, following the steps described above, accomplishing the ISO 14040 and 14044 guidelines and following ILCD Handbook recommendations. The aim is to show how the lightening of components in a vehicle can improve their environmental impact and therefore corroborate the benefits of the product developed. Different scenarios are being modeled in the study; these include a comparison of the environmental performance of the electric vehicle under study versus other vehicle technologies and fuels (diesel, petrol, LPG, CNG).
- *Recyclability Study*: This task of the project will compare the recycling of the plastic rack developed in Opera4FEV with the benchmark manufactured with aluminum. The life cycle thinking approach will be applied in order to avoid the generation of extra waste; different end of life scenarios for the plastic rack will be modeled.Recycling test of the thermoplastic selected will be implemented to

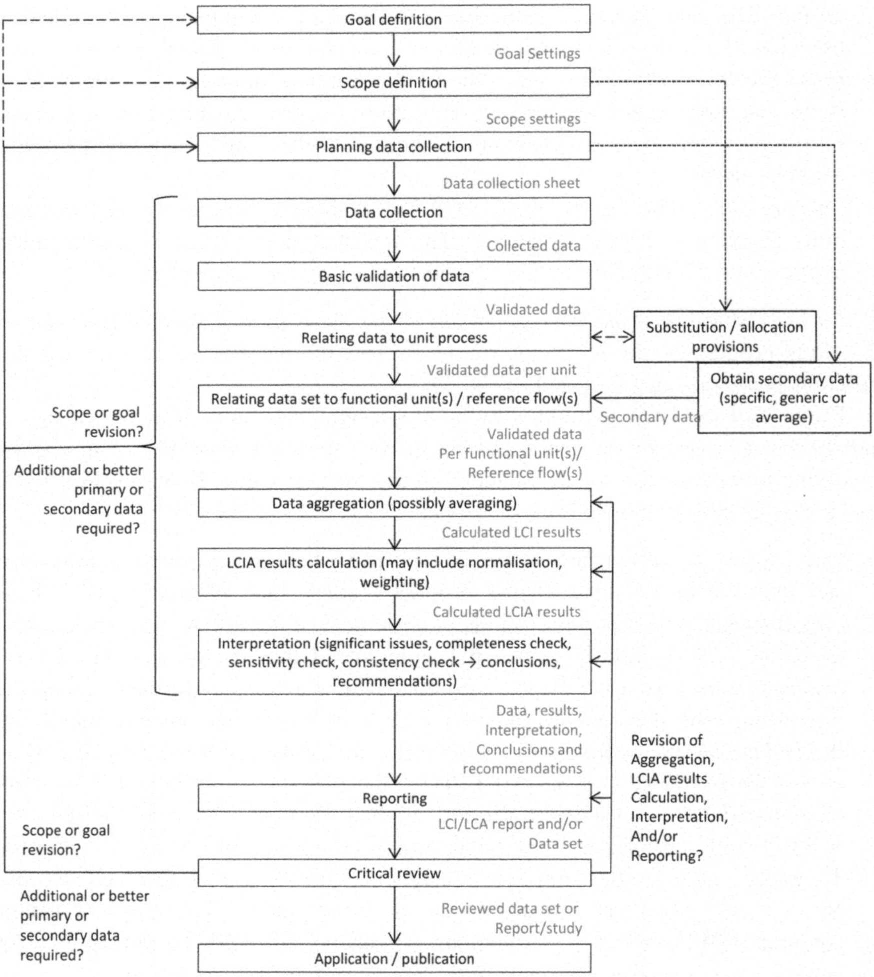

Fig. 2 Details of the iterative approach to LCA [7]

estimate how many times the material can be reused and reprocessed after its main use.

The Quick Eco scan has been carried for 4 different design options of the battery rack and has generated following results (Figs. 3 and 4):

Figures 3 and 4 include the impact of the manufacturing of the rack housing. The impacts of each design option include the manufacturing process of the different components of the rack system. Each component is made with a different material, except the Module parts, which are made of polypropylene (PP) with 20 % talc filling in all cases. In addition the impact of the injection molding process is included as well; and it aggregates the electricity, water and other emissions of the process. The order from higher to lower impact of the four options correlates in the

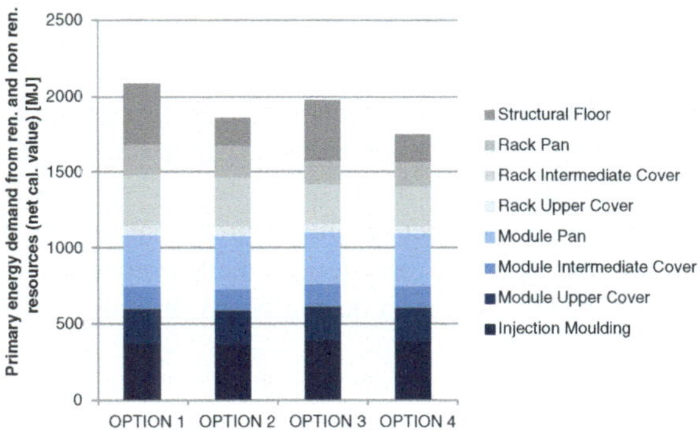

Fig. 3 Primary energy demand of the manufacturing of the four thermoplastic housing options

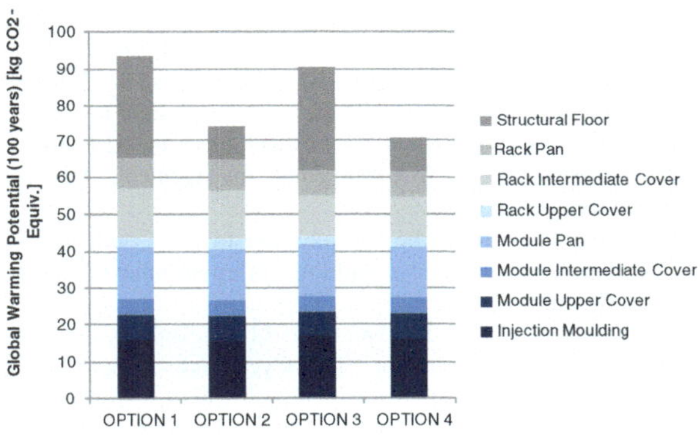

Fig. 4 Global warming potential of the manufacturing of the four thermoplastic housing options

two impact categories included here, namely Primary Energy Demand (PED) and Global Warming Potential (GWP) [8]. This impact assessment focuses on the structural floor and the rack pan, rack intermediate cover and rack upper cover, as these are the components that vary in the design options proposed for the study. The rack has two material options: acrylonitrile butadiene styrene/polycarbonate ABS/PC (options 1 and 2) and polypropylene (PP) with 30 % glass fibre (options 3 and 4). As it is shown in the figures, the PP has lower impact than ABS/PC. It can be noticed also that the material options for the structural floor have a large influence on the impact; polyamide (PA) with 50 % glass fibre has been used in options 1 and 3, PP with 50 % glass fibre in options 2 and 4. The manufacturing of the PA is more environmentally intense than the PP.

Based on the present results, combined with others included in the Quick Eco scan report (a sensitivity analysis has been included for a major scientific robustness in the decision making), design option 4 has been proposed as most optimal solution. Currently the full LCA and the recyclability studies are on-going in collaboration with different partners in the OPERA4FEV consortium.

5 Conclusions

The OPERA4FEV is an innovative European research project developing cost-effective innovative thermoplastic battery racks. It involves the complete value chain for current and future power battery systems racks for electric vehicles, and minimising the impact on the environment during the complete life cycle of the system by applying eco-design methodologies during the development of the system. In this context a design based on polypropylene has been proposed.

References

1. Van Mierlo J, Maggetto G, Lataire Ph (2006) Which energy source for road transport in the future? A comparison of battery, hybrid and fuel cell vehicles, energy conversion and management, vol 47, issue ECM-D-05-00636. ISBN-ISSN:0196-8904
2. Roadmap 2050: A practical guide to a prosperous, low-carbon Europe
3. A portfolio of power-trains for Europe: a fact-based analysis (2010)
4. ISO 14040 (2006) Environmental management—life cycle assessment—principles and framework, 1st edn
5. Communication from the Commission to the Council and the European Parliament (2003) Integrated product policy: building on environmental life-cycle thinking (COM (2003)302); Brussels 2003[1]
6. Sanfelix et al (2013) The enhanced LCA resources directory: a tool aimed at improving life cycle thinking practices. Int J Life Cycle Assess 18(1):273–277
7. European Commission—Joint Research Centre—Institute for Environment and Sustainability (2011) International reference life cycle data system (ILCD) handbook—recommendations for life cycle impact assessment in the European context. doi:10.2788/38479
8. Hischier R, Weidema B (2010) Implementation of life cycle impact assessment methods, ecoinvent report no. 3, July 2010

EASYBAT—Innovative Removal Battery Interfaces for Electric Vehicles

Chanan Gabay, Jacques Poillot and Yoav Heichal

Abstract New battery packs can make the EV more capable. Their share in the price of the Fully Electric Vehicle (FEV) is set to become even more dominant. Factors driving this include the strident demand for better car range. In addition, new battery packs increasingly incorporate electronics for safety and power conversion. The integration of these new complex battery packs presents major challenges to the industry especially considering the current lack of standards. The EASYBAT project, funded through the European Seventh Framework Program (FP7), will make it easier for European automobile and battery manufacturers to build EV with switchable batteries. EASYBAT will provide interfaces for switching a battery in and out of an electric car quickly and safely; the connector interfaces between the car, the battery, the communications network, and the battery cooling system; and design specifications that meet European industry and safety standards. The EASYBAT solution will be integrated and tested on fully electric vehicles to ensure it meets production-grade manufacturing criteria and European safety standards.

Keywords EASYBAT · EV · BSS · Battery switch station

C. Gabay (✉)
C.G - Smartech Ltd, 2 Hasharon Street, 4426931 Kfar Saba, Israel
e-mail: chgabay@gmail.com

J. Poillot
Renault, Paris, France
e-mail: jacques.poillot@renault.com

Y. Heichal
4 Hanarkis Street, 56905 Savyon, Israel
e-mail: Yoav.Heichal@gmail.com

E. Briec and B. Müller (eds.), *Electric Vehicle Batteries: Moving from Research towards Innovation*, Lecture Notes in Mobility,
DOI 10.1007/978-3-319-12706-4_6

1 Interfaces Requirements Specification

The EASYBAT system is built of 3 main actors: Vehicle, battery and the BSS (Battery Switch Station)—see Fig. 1. Each actor has at least 4 main interfaces: Mechanical interface, HV (electrical), LV (data) and Cooling interface.

1st task was to identify existing exchangeable battery interfaces solutions available on the market, analyze the overall improvements and modifications required to be done on the existing interfaces solutions, for both the vehicle and the battery necessary to develop a generic interfaces solution and requirements for the next generation interfaces solution.

The analysis of existing solutions and the constraints we identified regarding size, weight, height, tolerance and alignment lead us to conclude that the next generation battery type must be a "flat"/"pancake" battery located in the underfloor of the vehicle, and purely vertically switchable. This conclusion has a major impact on the solution and on the requirements.

1.1 Existing Solutions: On-Board Active Mechanism

In this solution, the vehicle plays an active role in releasing/locking the mechanical locks of the battery, while the battery switch station is not involved in the mechanical locking mechanism, only accepts the released battery and carries it into the station storing/charging area. Better Place demonstrates this concept on various occasions such as Japan's METI EV taxi program.

This concept based on mechanical latches/locks located in the underbody of the vehicle that are responsible for the following functions:

- securing the battery pack to vehicle body
- aligning the battery pack with vehicle body during battery installation
- preloading the battery pack against the vehicle body.

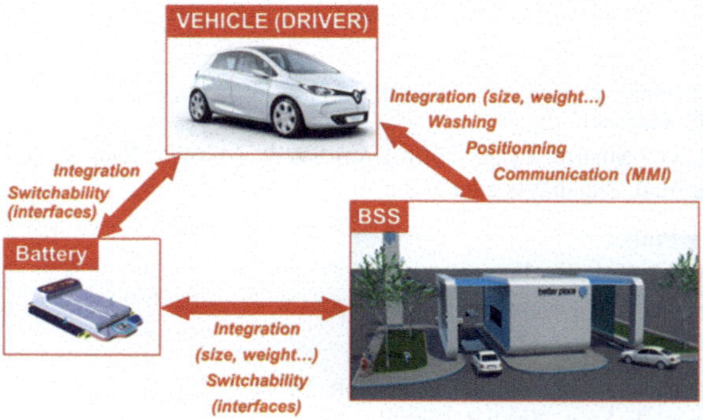

Fig. 1 EASYBAT main actors: vehicle, battery and the BSS

1.2 Existing Solutions: On-Board Passive Mechanism

In this solution, developed by Renault, the battery switch station plays an active role in releasing/locking the mechanical locks of the battery, while the vehicle is not involved in the locking/unlocking control. This solution concept is implemented commercially in the Renault Fluence ZE, and is based a "toolbox" carried below the vehicle (by the BSS robot) which locks/unlocks the battery.

1.3 Requirements Identification

EASYBAT requirements are the result of analysis done on the current solutions. We identify the advantages and disadvantages, and the improvements needed for the next generation solution, while for a question of simplicity of the process (of both BSS and assembly plant) the battery shall be assembled in the vehicle vertically.

1.3.1 Vehicle Requirements

Battery trunk location is favorable for the vehicle height but not for the trunk capacity and functionality. EASYBAT requirement is to keep trunk capacity and functionality optimization. Therefore, the proposed battery location is under the body of the vehicle (like in the case of ZOE architecture). In this case, the vehicle critical paths are made by the ground clearance, mechanism and/or the battery vertical dimension, floor structure thickness, driver and/or passenger position (heel location and ergonomic requirements) and head clearance versus vehicle height (see Fig. 2).

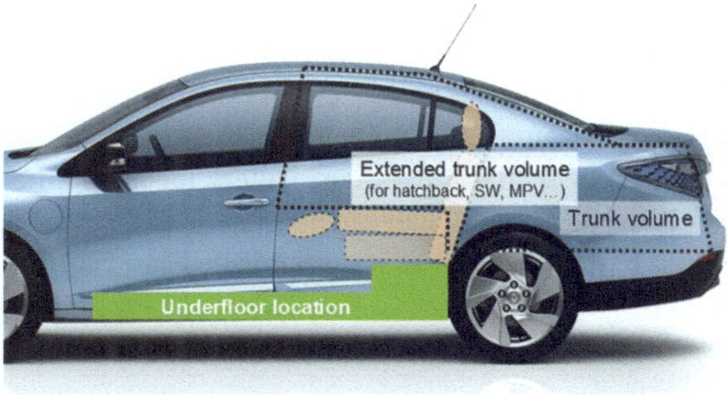

Fig. 2 Underfloor location—trunk volume and functionality

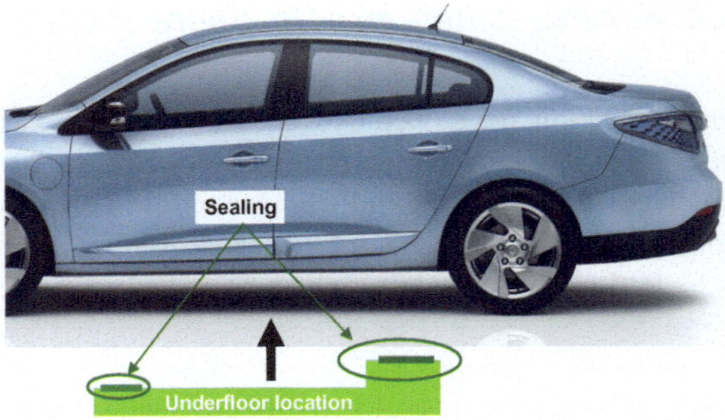

Fig. 3 Electrical and thermal sealing requirement (dust and water proofness) and location

1.3.2 Interfaces Requirements

Electrical and thermal interfaces shall be dust and waterproof. So sealing is needed to protect them and to protect the battery against dust and water penetration. In order to simplify the BSS interfaces, the sealing will be located on the battery side rather than on the vehicle/BSS side (see Fig. 3).

Mechanical interfaces dimensions shall be such as they will not impact battery modules in application of all standard crash tests (regulation and euroNCAP), and in particular all kind of lateral pole crash (same test speed as regulation).

The electrical interface is already developed and standard. Its definition shall be carried over. However, it should be cost reduced.

1.3.3 Battery Requirements

The air or liquid cooling parameters shall respect the specified values in order to keep cooling efficiency regarding battery performance and durability in one hand, and battery internal components reliability in other hand.

1.3.4 Requirements Analysis Conclusion

All the above analysis of current solutions and the constraints we identified regarding size/weight/height lead us to conclude that the next generation battery type must be a "flat"/"pancake" battery located in the underfloor of the vehicle, and purely vertically switchable. This conclusion has a major impact on the solution and on the requirements that were published in [1].

1.4 System Architecture

The architecture identifies the external and internal interactions points of a switchable battery. It will fulfil the functional and the non-functional requirements to provide a safe, reliable, extensible, cost effective and open platform for handling all interactions between the EV battery pack and the vehicle.

The architecture emphasis specifically the mechanical mechanism, that base on the requirements for tolerance and alignment, are very difficult to implement (from technological, time and cost points of view), making a mass-market implementation of the current solution very expensive. Therefore, one of the leading motivations for coming up with a new architecture concept (see Fig. 4), is to simplify and improve the alignment process.

For that, we defined the location, dimension and actuation of the mechanical mechanism to serve as reference to the position of all other connectors (electrical, data and thermal).

One major issue was the requirement, to define a generic solution suitable for all vehicles, from the small 2-seat A class through to the D class, taking into consideration the constraints on the switching station. It results in a generic architecture concept whose implementations were divided into 2 solutions. One solution meet the requirements for the small A class vehicles, the second solution meet the requirements for the larger B, C and D classes.

The aim of the new solution is to get the best compromise between the existing solutions in order to cumulate as much as possible their advantages while limiting their inconveniences. The evolution concerns only the mechanical interface and the

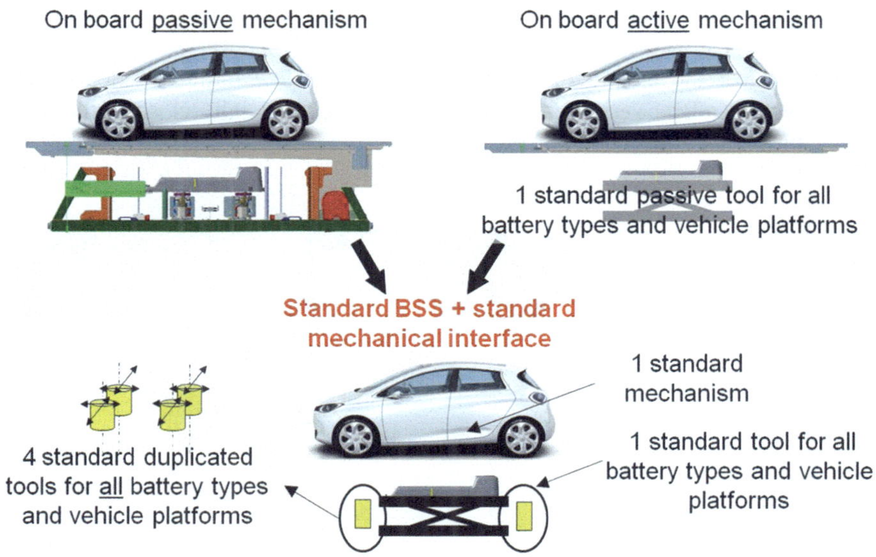

Fig. 4 The new EASYBAT architecture

way to control it. The principle is to replace the tool-box, the centering and attitude positioning elements currently used for the on-board passive mechanism solution, by 4 independent duplicated compliant modules using standard tools and able to reach each standard mechanical interfaces independently in XYZ and rotation θZ position. Their flexibility allows managing different vehicle and battery types. In this case, this equipment is to be invested only one time. On the other hand, it facilitates standard and low cost battery conveying equipment.

1.5 Generic Mechanical Battery Interfaces

As mention before, the main EASYBAT solution interface is the mechanical interface that should assure battery unlocking/locking to the vehicle body in a way that allowed a purely vertical switching taking care of any battery to vehicle mis-alignment, in a reliable and safe way.

This should simultaneously allow the electrical vertical disconnection/ connection.

At the same time, the solution should be able to integrate with current vehicle architecture environment, for example the current ZOE using a standard and multidirectional tool.

The mechanical interface (see Fig. 5) is comprised of a latching device in the shape of a rotating hook. By rotating the mechanism spindle screw, using the BSS module, the spindle translates a nut, which moves in a vertical direction. The nut of the mechanism coupled to a push rod and bell-crank assembly. The bell-crank attached rigidly to the main hook, when nut is moving vertically, the motion of the nut causing a rotation of the hook between "open" and "close" positions. Due to the kinematics of the mechanism, the rotation of the spindle has very high mechanical advantage over the motion of the hook; this means low torque exerted on the spindle screw inflicts very high torque on the hook, especially close to the hook end-of-stroke, near the hook's "close" position.

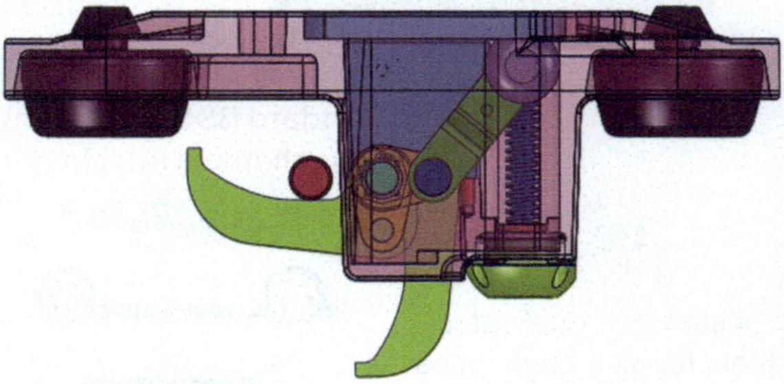

Fig. 5 EASYBAT mechanical interface

2 Conclusions

EASYBAT project achieved its main goal, develop generic interfaces for mechanical, cooling, and data and electrical systems between the Electric Vehicle (EV), Battery and Battery Switch Station (BSS).

We believe that all EASYBAT interfaces can be applied in a generic way to other EV's and Hybrid EV's (HEV's), and for each interface type, the OEM can attempt to use them 'as is' or with some minor modifications if required.

Reference

1. EASYBAT deliverable 1.1 (intermediate interfaces requirements specification), 2012

SuperLIB: Smart Battery Management of a Dual Cell Architecture for Electric Vehicles

Can Kurtulus, Peter Krabb, Volker Hennige, Mika Räsänen, Justin Salminen, Matti Nuutinen, Joschua Grosch, Michael Jank, Erik Teuber, Vincent Lorentz, Martin Petit, Joseph Martin, Jean-Louis Silvi, Noshin Omar and Dhammika Widanage

Abstract A dual-cell battery concept has been proposed to address electro-mobility challenges where the concept entails a combination of high energy and high power optimized cells combined with an advanced management system. This concept and its advanced management system are being investigated as part of a European Seventh Framework Programme research project with the name SuperLIB, where the target is to extend life time of the battery and utilize an advanced battery management system to increase overall performance. An overview is provided on the

C. Kurtulus (✉) · P. Krabb · V. Hennige
AVL List GMBH, Hans-List Platz 1, 8020 Graz, Austria
e-mail: can.kurtulus@avl.com

P. Krabb
e-mail: peter.krabb@avl.com

V. Hennige
e-mail: volker.hennige@avl.com

M. Räsänen · J. Salminen · M. Nuutinen
European Batteries, Karapellontie 11, 02610 Espoo, Finland
e-mail: mika.rasanen@europeanbatteries.com

J. Salminen
e-mail: justin.salminen@europeanbatteries.com

M. Nuutinen
e-mail: matti.nuutinen@europeanbatteries.com

J. Grosch · M. Jank · E. Teuber · V. Lorentz
Fraunhofer IISB, Schottkystraße 10, 91058 Erlangen, Germany
e-mail: joshua.grosch@iisb.fraunhofer.de

M. Jank
e-mail: michael.jank@iisb.fraunhofer.de

E. Teuber
e-mail: erik.teuber@iisb.fraunhofer.de

V. Lorentz
e-mail: vincent.lorentz@iisb.fraunhofer.de

E. Briec and B. Müller (eds.), *Electric Vehicle Batteries: Moving from Research towards Innovation*, Lecture Notes in Mobility,
DOI 10.1007/978-3-319-12706-4_7

main areas of development including cell design, battery management system development with advanced algorithms and energy distribution and advanced temperature sensor development.

Keywords Battery management · Energy distribution · Li-ion battery · Dual cell · Printed sensor · Temperature sensor · Electrical modeling · Thermal modeling

1 Introduction

Energy storage system is the most critical component for electric vehicles, and performance, cost, durability and reliability of this component will determine how widespread electric vehicles will be used. A dual cell concept has been proposed where a combination of energy optimized and power optimized Li-ion battery cells, managed by an advanced battery management system, targets an improvement in all the aforementioned areas.

The proposed concept is being explored as part of a European Framework 7 Programme research project called SuperLIB [1], where technology development includes an advanced battery management system with energy distribution algorithms, a novel thermal sensor for close monitoring of each cell, thermal and electrical modeling to aid in the advanced battery management system development and mechanical design, development of the power and energy optimized cells, as well as an integrated mechanical design concept.

The envisioned technological advances aim to accomplish a multitude of objectives, such as extending the life time of the battery, while meeting power and energy requirements of electric vehicles at the same time. This is accomplished via using an extended SoC range of the energy optimized cells—due to the fact that

M. Petit · J. Martin
IFP Energies Nouvelles, Rond-point de l'échangeur de Solaize – BP 3, 69300 Solaize, France
e-mail: martin.petit@ifpen.fr

J. Martin
e-mail: joseph.martin@ifpen.fr

J.-L. Silvi
Valeo EEM, 2 Rue André Boulle, 94046 Créteil Cedex, France
e-mail: jean-louis.silvi@valeo.com

N. Omar · D. Widanage
Vrije Universiteit Brussel, Pleinlaan 2, 1050 Brussels, Belgium
e-mail: noshomar@vub.ac.be

D. Widanage
e-mail: wwidanag@vub.ac.be

Table 1 Pack level system specifications

	SuperLIB
Specific energy (Wh/kg)	>75
Energy density (Wh/L)	>125
Peak discharge power (kW), 10 s	90
Peak charge power (kW), 10 s	75
Usable energy (kWh, EoL)	15

power requirements are no longer relevant for them, whilst covering power requirements via the power optimized cells. In addition, since high current rates are handled by the power optimized cell, the stress on the energy cell is greatly reduced, with an accompanying extension of life time.

System performance targets were based on energy intensive traction battery pack requirements such as PHEVs and EVs. These applications require a relatively large battery pack, and substantial power capabilities. Specifications of the SuperLIB battery pack were defined to cover a large range of performance needs, and can be found listed below in Table 1.

The paper is structured to first discuss development of the cells, continue with modeling and characterization, then delve into novel temperature sensor development and finally depict the overall management concept.

2 Cell Development

As part of SuperLIB, the target was to design and manufacture both (a) High Energy (HE) Cells, and (b) High Power (HP) Cells and analyze their properties in the Dual-Cell Battery system. Through the knowledge of the performance and operational conditions, one can make intelligent estimates of the performance or the actual Dual-Cell system. In this analysis, we demonstrate that the battery cycle-life performance should not only be defined as the ability to maintain battery Ah-capacity, but energy-roundtrip efficiency, i.e. charge/discharge energy efficiency, is a more valid indicator of battery cycle-life performance. The battery should be used with charge and discharge rates and within a temperature range, where the round-trip energy efficiency remains constant during the battery life. The ability to maintain constant roundtrip efficiency is also a good indicator of good battery quality. If round-trip efficiency degrades during battery life, available energy content of the battery is decreased due to the increasing amount of wasted energy turning into heat.

Present day lithium ion batteries have limitations, but significant improvements have recently been achieved. The main challenges and improvement areas of lithium ion batteries are related to (a) material deterioration, (b) operating temperature range, (c) energy and (d) power output, and (e) lifetime.

Table 2 Specification of the HE and HP cells used in the project

Cell characteristics	HE cells	HP cell
Specific energy (Wh/kg)	145	130
Rated capacity (Ah)	45	24
Nominal discharge voltage (at C/3) (V)	3.2	3.2
Width (mm)	165	165
Height (mm)	275	275
Thickness (mm)	13	7
Maximum discharge current (A)	180	216
Maximum charge current (A)	45	81

Note that all values are for 23 °C

Increased lifetime combined with a higher recycling rate of battery materials is essential for the electric vehicle industry. There are a number of options involving different chemical processes and battery types.

The cathode battery chemistry used in this project is lithium iron phosphate ($LiFePO_4$), which has shown good thermal, cycle and storage life performance. Both the designed and manufactured HP and HE cells share the same $LiFePO_4$ cathode material, but have otherwise other materials and different cell design. Both cells have carbon anode, but the composition of the anode is different due to the different design requirements. Short specifications and characteristics of HE and HP cells are summarized below in Table 2. The values of Table 2 are used as a starting point in the HE and HP cell design.

As seen from Table 2, the main difference between the cells is the difference in allowed discharge and charge current/power. As an example, Fig. 1 summarizes the different discharge behavior of the cells. It can be seen clearly from the figure that the HP cell can be discharged with nearly double the C-rate of the corresponding HE cell. It should however be noted that the 8C discharge current of the HP cell is approximately the same current as the 4C discharge of the HE cell. Moreover, the temperature rise behavior is about the same with the same current. The HE cells can, of course, provide much more energy.

The discharge and charge performance parameters do not unfortunately indicate the complete performance profile of the battery to be used in an electric vehicle (EV) application. The thermal performance or storage performance and more importantly the cycle performance have the largest effect on the overall performance of the EV in practical use. Here we concentrate on cycle performance, since the thermal/storage performance is also affected by the thermal management system. This thermal management system also affects the battery cycle-life, but these effects can be taken into account as a corrective term in the cycle-life model. In any case, cycle-life data need to be collected for an understanding of the battery cycle-life performance.

Cycle life of HE and HP cells is shown in Fig. 2, where HE cells have been charged and discharged with 1C/1C and 2C/2C-charge/discharge cycle and HP cells

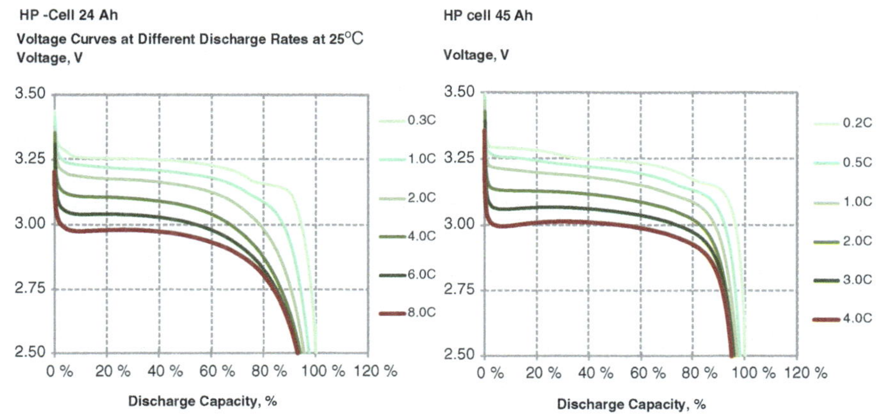

HP Cell performace (ambient 23°C)

C-rate	Capacity, Ah	Average Voltage, V	Energy, Wh	Max temp,°C
0.3	23.2	3.20	74	23.4
1.0	22.7	3.15	71	27.2
2.0	22.2	3.09	69	31.2
4.0	21.9	3.01	66	36.3
6.0	21.7	2.96	64	40.0
8.0	21.6	2.91	63	41.9

HE Cell performance (ambient 25°C)

C-rate	Capacity, Ah	Average Voltage, V	Energy, Wh	Max temp,°C
0.2	46.1	3.22	148	25.0
0.3	45.7	3.20	146	25.8
0.5	45.3	3.18	144	27.1
1.0	44.7	3.13	140	30.5
2.0	44.3	3.07	136	36.8
3.0	44.1	3.02	133	40.6
4.0	43.8	2.98	130	44.3

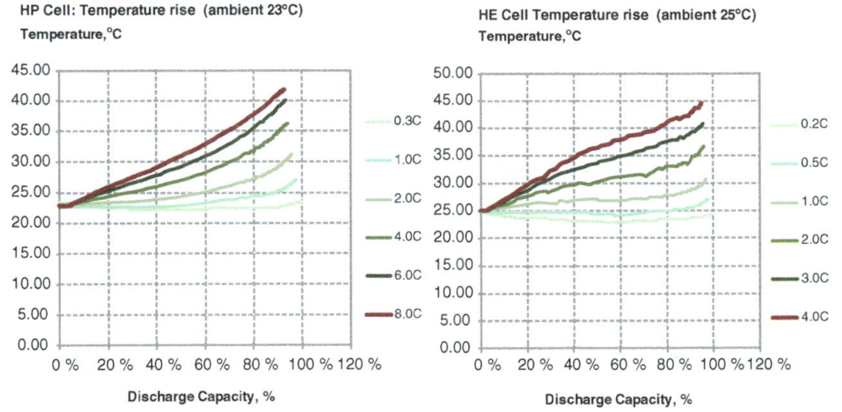

Fig. 1 HP and HE cell performance in 100 % depth-of-discharge with different discharge currents. The charging method is constant current/constant voltage: using 22.5 A current with 3.55 V voltage limit and end-condition current 0.03C (1.35 A). Discharge cut-off voltage is 2.5 V

with 2C/2C and 3C/3C cycle. Figure 2 indicates that the HE cells have approximately the same cycle-life performance with 1C/1C and 2C/2C charge/discharge and HP cells have the same performance with 2C/2C and 3C/3C charge discharge. However, when we look at the roundtrip energy efficiency, it becomes evident that 1C charge/discharge rate is a better choice for the HE cells, as opposed to the case considering capacity degradation only.

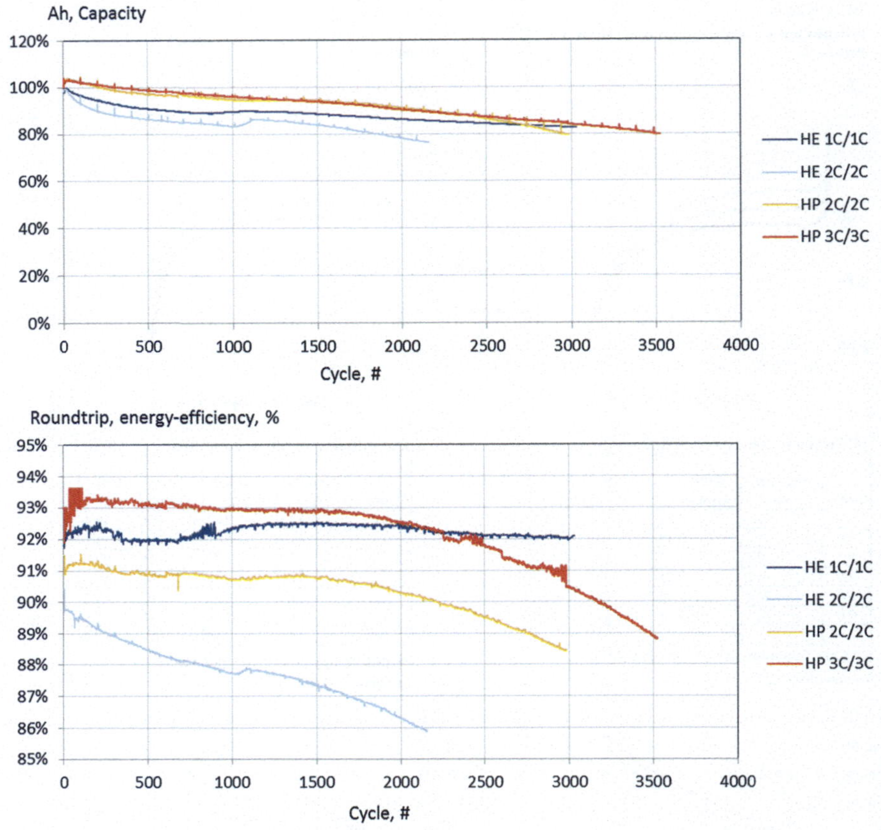

Fig. 2 HP and HE cell cycle-life: Ah—capacity and round-trip energy efficiency in 100 % depth-of-discharge with 1C/1C, 2C/2C and 3C/3C charge discharge cycles

3 Battery Modeling

3.1 Electrical Modeling

In the context of development of a battery management system, a battery model can be considered as a key issue. Therefore the development of accurate electrical and thermal models are of high importance. In the framework of this project, Vrije Universiteit Brussel has developed an electrical battery model (see Fig. 3) for the HP and HE cells in Matlab/Simulink. The performed investigation revealed that charging and discharging of the battery may not be assumed only by a simple Ohmic resistance and an open circuit voltage source. To model the short-term (a few seconds) dynamic behavior, an Ohmic resistor and a RC circuit along with a series capacitor is required. Then, the model has been extended whereby the charge and discharge effect has been included in the model as presented in Fig. 4.

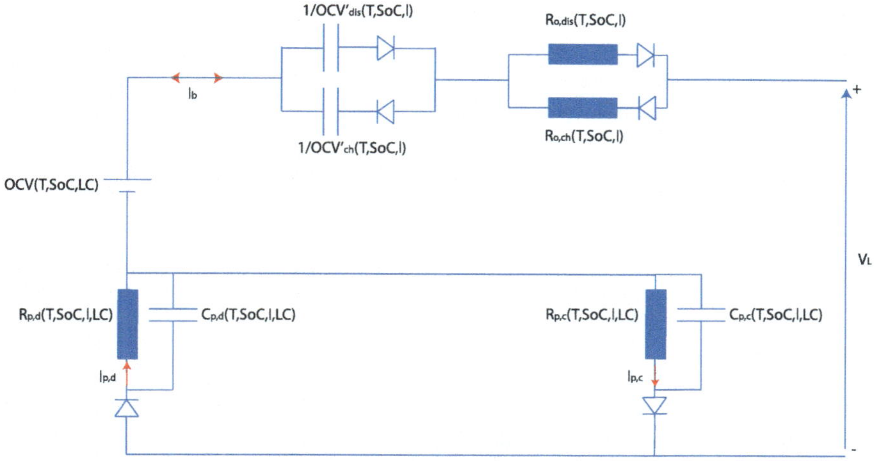

Fig. 3 SuperLIB battery model

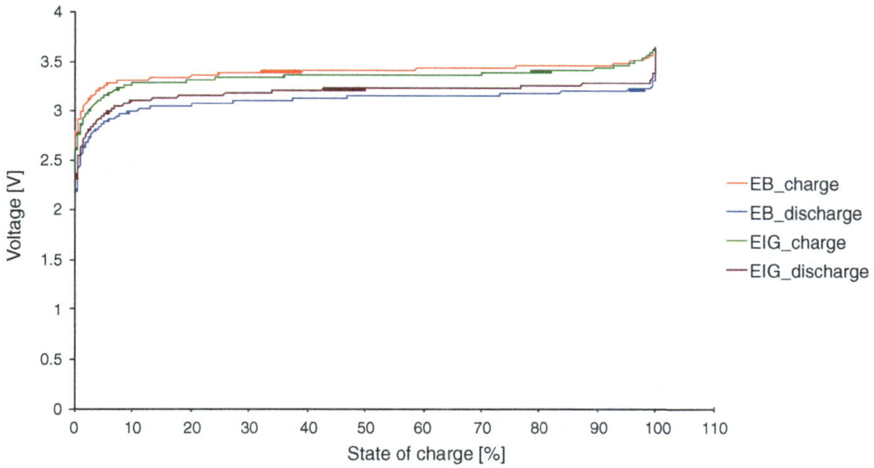

Fig. 4 1C charge and discharge evolution of the HE and HP cells

The main conclusions of the HP and HE electrical models are that the simulation performance is more accurate, based on the percentage relative error, over a large SoC window at the two higher temperatures of 25 and 40 °C; whereas at 0 and 10 °C the accuracy of the simulation performance reduces giving in general a relative error of less than 12 % over a reduced SoC window of 20–85 %. Furthermore, at the two low temperatures of 0 and 10 °C, the HP model simulation accuracy is better since its relative error percentage is lower (10 % at 0 °C and 5 % at 10 °C) compared to the HE model (12 % at 0 °C and 10 % at 10 °C) over the SoC window of 20–85 %. A summary of the relative error of the HP and HE model over

Table 3 Relative error of the HP and HE models

Temperature (°C)	HP model relative error (SoC window)	HE model relative error (SoC window)
0	Less than 10 % (over SoC of 10–85 %)	Less than 12 % (over SoC of 20–85 %)
10	Less than 5 % (over SoC of 10–85 %)	Less than 10 % (over SoC of 20–85 %)
25	Less than 5 % (over SoC of 5–95 %)	Less than 10 % (over SoC of 8–93 %)
40	Less than 5 % (over SoC of 5–98 %)	Less than 5 % (over SoC of 4–99 %)

a certain SoC window and at the four temperatures is tabulated in Table 3. The relative errors listed in Table 3 are defined based on the allowed voltage swing of a cell, so a relative error of e.g. 5 % would correspond to an absolute error of %5 × $(V_{max} - V_{min})$.

Here it should be underlined that the higher error percentage for the HE cells is due to the fact that the cells are prototypes cells and not very well optimized.

3.2 Thermal Modeling

In order to assess and simulate the electro-thermal behaviour of the battery cells used in the SuperLIB project, two different levels of models have been developed calibrated and validated at IFPEN for both HE and HP cells.

The quasi-static (QS) model is able to simulate the thermodynamics of the system as well as the internal resistance but does not take into account the diffusion and other transient phenomena. The dynamic model is able to take these transient phenomena into account but needs a more complicated calibration. The electro-thermal models compute the cell voltage and the generated heat flow and are then coupled to a thermal model that will compute the skin temperature.

Once calibrated, both models have been validated against experimental data simulating typical usage of battery cells (constant current charges and discharges and HEV duty cycles). Some results of HE cells and HP cells are shown in Figs. 5 and 6.

The validation showed that the models are able to simulate very precisely the thermal behaviour of both cells at temperature ranging from 0 to 40 °C (error ±1 °C). Concerning the voltage prediction, the lower the temperature is, the bigger the error gets. However for temperatures above 10 °C error on voltage remains below 50 mV for the dynamic model in HEV duty cycles.

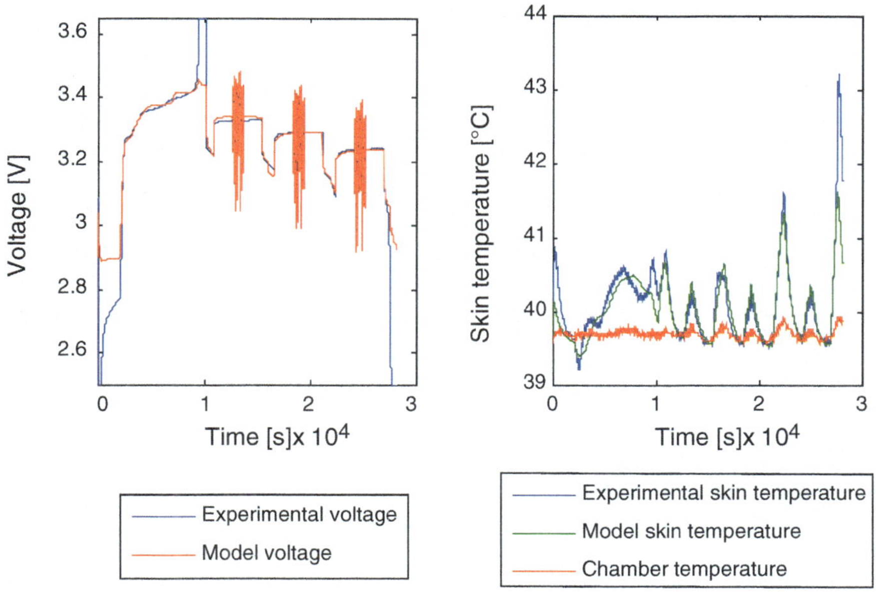

Fig. 5 HE cell model validation at 40 °C on a HEV duty cycle

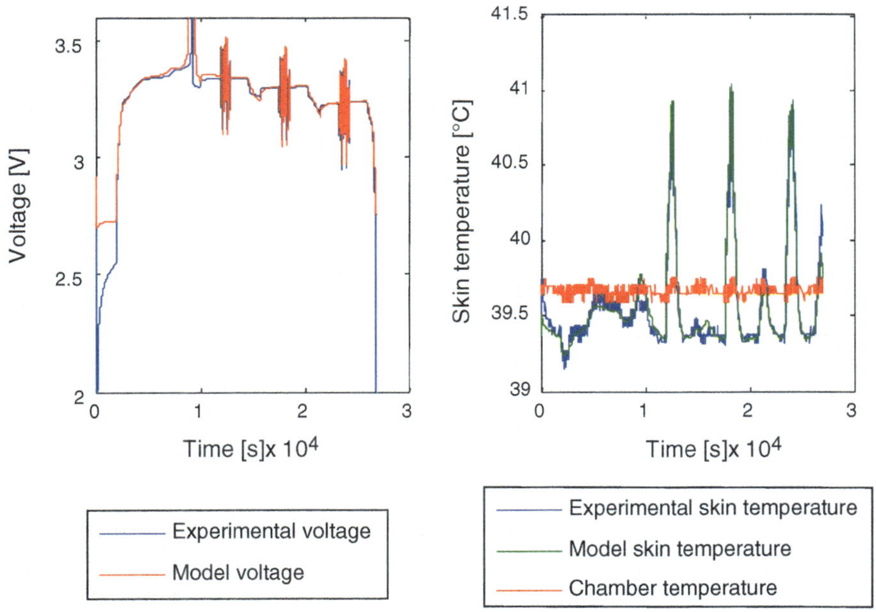

Fig. 6 HP cell model validation at 40 °C on a HEV duty cycle

4 Advanced Temperature Sensor Design

The concept of the temperature sensors used in traction battery packs has to be as simple as possible. This ensures cost efficient and highly reliable devices, which are easy to handle in high volume production.

Another point is the demand for mechanical flexibility of the devices, as well as the substrate. The devices have to withstand mechanical forces due to expansion and contraction of the battery cells during operation. These single battery cells can be damaged by commercial sensors with a pearl-shaped package. A flat and thin sensor with a large surface area, that consumes little mounting space, is a prerequisite for a temperature sensor to be placed directly onto the battery cells, or else in between neighboring ones. The aforementioned points lead to the following definition of sensor requirements by Fraunhofer IISB:

- Low-cost processing,
- Thin, flexible architecture,
- Layout customizable to different cell designs,
- High-TCR material with high noise tolerance.

Printed electronic technologies meet these requirements to a high extent and are also suitable for high-volume production, for example with sheet-fed or roll-to-roll processing, at high cost efficiency.

The demand for thin and flexible substrates imposes an additional requirement: as most of the suited substrates like plastics or paper are sensitive to temperature, processing should be done at a low thermal budget, i.e. in the range below 150 °C. Furthermore, thin printed layers constituting the temperature sensor yield a low heat capacity and therefore result in a low time delay for the detection of temperature changes of the battery cells. The low thickness of the printed elements (compared to conventional discrete temperature sensors) allows an easy placement between the battery cells. Moreover, the placement of the sensor onto the cell surface, or even inside the cell packaging, is possible.

The repartition of multiple devices (e.g. in a matrix structure) over the complete surface of the battery cells is made possible. Using a sensor matrix, the temperature distribution over the cell can be determined, thus ensuring high performance and safety regarding the early detection of possible malfunctions of the battery cell (e.g. thermal runaway).

The layout of the temperature sensor is shown in Fig. 7. The sensor is realized on a plastic substrate. Processing starts with a screen printed layer of silver paste, which yields the wiring and contact metallization. On top, the temperature sensitive layer is deposited, again by screen printing. The sensor fabrication is finalized by application of encapsulating layers that are intended to shield the sensor from moisture, contaminations and insulating it electrically. The chosen sensor architecture delivers low complexity, high robustness and can be integrated with complementary devices on the same technology for the realization of more complex sensor/actor and control solutions.

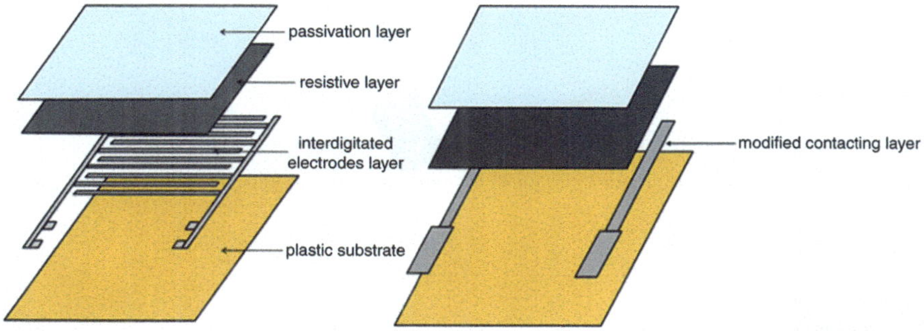

Fig. 7 The *left* figure shows the general device layout of the temperature sensor, cell balancing resistor, or cell heating element. The *right* figure shows the adjusted structure for the temperature sensor. The resistance of the sensor was adjusted by modifying the W/L ratio of the sensitive layer

The adjustable parameter in the layout is the ratio of width and length of the resistive element (W/L ratio). Depending on the resistance ranges of the temperature sensitive material and the resistivity of the wiring, the W/L ratio is customized to meet the requirements of the read-out electronics and the desired areal resolution of the sensing spot. Spatial inhomogeneities in the material parameters of the temperature sensitive material can be compensated without influences on the general system by enlarging the W and L values while keeping the W/L ratio constant. On the other hand, reducing W and L leads to a higher spatial resolution of the temperature measurement. The simplest structure for sensor design is a rectangular sensitive feature with contacts on two opposite sides of the sensitive element. With high resistivity materials, the W/L ratio can be increased significantly by interdigitating fork-like fingers alternatingly consisting of the two electrodes. In the opposite case, the sensor layout for sensitive material with extremely low resistivity can comprise a narrow line. Length maximization within a given area can be realized by meander structures.

The general device layout can be utilized for a range of passive device concepts that may deliver additional functions of a modular BMS system, e.g. for balancing resistors, heaters, or chemical sensors for humidity measurement. The base function of these additional devices can be realized by replacing the sensitive layer by an alternative functional material and adapting the layout.

Regarding the functionality of a balancing resistor, the electrical energy of the balancing process is converted directly into thermal energy, which also can be used to reach an adequate working temperature level of the battery cell in cold winter conditions. Furthermore, it is possible to homogeneously heat up the battery pack (e.g. in cold-start situations or to enable charging in winter conditions) by heating each single battery cell directly at its core, without any additional expensive external heating source assembly. This will increase the overall battery cell performance as the heating energy dissipation to the module environment and other mechanical parts (e.g. module package frame) can be reduced to a minimum.

Within SuperLIB the focus of development is on the cell temperature sensor (Fig. 8). Details about the choice for the functional layers are given below.

Fig. 8 Photographs of the interdigitated electrodes layer (*left* picture) and the completely printed temperature sensor developed within the SuperLIB project (*right* picture)

The sensitive material in our case is a polymer filled with graphite particles. The principle of the temperature-induced change of resistance is the expansion and shrinking of the polymer with increasing/decreasing temperature. With expansion of the polymer, the graphite particles are separated from each other. This may lead to a reduction of interparticle contacts and result in an increasing resistance due to the loss of percolation paths along the current flow direction. For shrinking polymer with decreasing temperatures the principle applies vice versa.

The wiring is used simultaneously for contacting the sensitive element of the sensor. For the experiments, we utilized an ink jet ink commercially available by Sun Chemical as well as a low-temperature curable screen printing paste by DuPont (DuPont 5064H). The screen printing paste delivers a thicker silver layer and a considerably lower defect density. Due to the long wiring to the external connector, the screen printing paste was chosen for deposition of the metal layer in the final devices. Silver inks are the current industry standard for printed conductor materials. However, due to the expensive raw material, alternative materials like carbon, copper or aluminum are under investigation and are expected to enable a low-cost replacement of silver in the future.

The encapsulation of the sensor was realized with a range of polymeric materials known from semiconductor thin-film technology or electronics technologies (Microchem SU-8 2000, PMMA 2k, Plastik 70 spray encapsulant, Polyurethane IsoPur, and conventional lamination bags). Due to the best performance in the device application, IsoPur and lamination bags were chosen in depth investigations. The lamination process with a foil leads to a thicker overall device but a good and tight passivation. Lamination is performed using an office-type lamination device. The polyurethane (IsoPur K760) is applied by doctor blading but can also be processed by screen printing if needed.

The sensor data shown in Fig. 9 were recorded in a standardized temperature cycle (+25 to +65 °C) that covers the most important operation range with respect to safety issues and cell wear-out due to cell cycling. Extended characterization

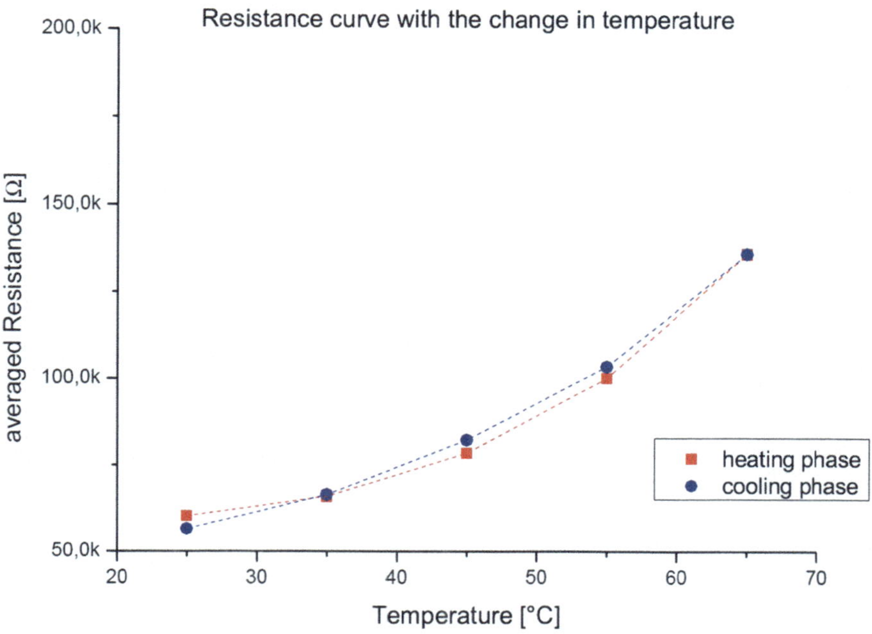

Fig. 9 Electric resistance of the sensor over the temperature

between −25 and +65 °C shows that the temperature sensor offers the best sensitivity above room temperature with a strong reduction of the TCR below +10 °C. As the operation of the cells at very low temperatures can be handled by the higher-level battery monitoring system, the low temperature resolution in the low-temperature region can be accepted. In this regime, the sensor can be utilized as an effective cell-based monitoring device for detection of a predefined "shut-off" or "turn-on" temperature.

To summarize the results of the temperature sensor device, the basic requirements like low-cost manufacturing, thin form factor, low effort for device assembly, and low read-out complexity were met by a thermistor concept based on a plastic substrate and screen printed functional layers.

The future perspective of the sensor placement is the integration of the printed sensors directly on parts or inside of the encasement of the battery cell. By increasing the number of sensors a sensor array could be assembled which makes it possible to get detailed information about the temperature distribution on the surface of every individual battery cell. This array could also be applied for heating the battery. Also, the device concept is open for the integration of additional or complementary materials for the realization of additional functionalities (e.g. chemical sensors) or the improvement of device characteristics (temperature ranges, sensitivity) depending on the respective application case.

5 Battery Management System

5.1 Electrical Architecture Description

In the SuperLIB application, the high energy storage component is made up of 45 Ah cells, and the high power storage element consists of 7 Ah cells where each HE module will have 14S configuration and each HP module will have 14S3P configuration. The battery pack will be composed of seven (7) HE modules in serial and seven (7) HP modules in serial.

This choice of 14S cells module allows making module with voltage compatible with VDA standard. It also presents the advantage to improve the synergy with European funded project ESTRELIA through the integration of 2 AS8506 in daisy chain in order to manage up to 14 cells.

For the demonstration need, the number of cell temperature sensors will be 14 per module.

The following elements have led to the choice of making a connection at pack level instead of module level:

- We only need one contactor per string and one contactor with preload on HV side
- We only need 3 current sensors (1 per string + 1 for the external current)
- We only need 2 Fuses (1 per string).

Making a connection at module level would multiply by 7 all these elements. It also multiplies by 7 the number of unreferenced DC to DC converter and Battery Control Units.

The choice of having 1 Module Control Unit per HE and HP module with cell voltage acquisition, temperature acquisitions and balancing circuit presents the following advantages:

- Easier integration in the module.
- Shorter wire length.

The following section presents the functional block diagram of the system at battery level and at the BCU level (see Fig. 10).

This slave master architecture with connection at pack level (voltage level of 360 V) was chosen according to the considerations presented previously.

The Battery Control Unit integrates the main processor which will calculate the battery parameters (SoC, SoH, SoF), drive the contactors and diagnose the global function of the battery. It also hosts the converter link drivers and is directly linked to the power electronics.

The Module Control Unit manages one HE module or one HP module with cell voltage acquisition, temperature acquisition and balancing circuits.

The balancing of HE cells and HP cells is independent and supports both passive and active applications. The layout integrates both possibilities and assembly of the boards is different according to the choice of active or passive balancing.

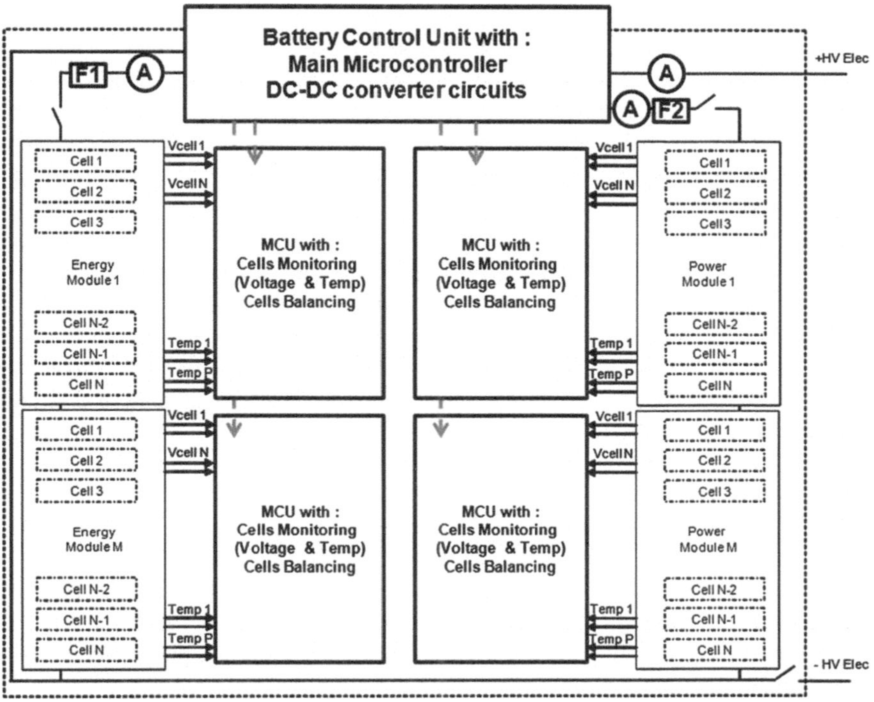

Fig. 10 Functional block diagram at battery level

The electronics does not integrate any driver for external cooling actuators. These actuators are driven through a CAN message.

As a safety measure, a protection scheme linked to the board-to-board communication is integrated and the MCU has the ability to open the main contactors of the battery in case of a problem.

5.2 Energy Distribution Concept

The BMS software is divided into different software parts. This modularity ensures a high flexibility for adaptation of certain parts with minor effect on the overall software. The part which controls the energy distribution between the High Power (HP) and High Energy (HE) string of the pack is the HP/HE Energy Flow Management. This control strategy on one hand prevents the HE string from stress situation, i.e. high currents which reduces the cycle lifetime, and on the other hand keeps the HP part on an optimal SoC level as it is done for HEV battery packs. The latter is important to provide sufficient power in situations such as sudden power demand and regenerative breaking events without stressing the HE part. Furthermore with this concept the usable SoC range of the HE side can be increased (5–95 %) without power limitations or sacrificing the battery life time.

The HP/HE energy flow management is a rule based control strategy, which manages the energy distribution taking into consideration battery temperature, SoC levels of HP and HE string and vehicle demand. This is done by adjusting a current set point at the DC/DC converter which limits the current on the HE side. Since the energy content of the HP string is relatively high, the function also provides an operation mode for charge depletion of the HP string in order to use a greater amount of the stored energy and improve the electric range. The depletion of the HP SoC is activated if the HE SoC is below a certain level.

The basic functionality of the energy flow management can be seen in Fig. 11.

The set point at the DC/DC converter is first adjusted depending on the actual temperature of the HE string. The energy distribution concept can be classified into different working conditions;

5.2.1 Light Load Condition

If the SoC of the HP string is within a desired window, the HE side provides the main power. Additionally if the HP SoC is below this window, the HP string is charged from the HE string. In case the HP SoC is above the optimal window, only the HP string is used.

5.2.2 Heavy Load Condition

In case of high vehicle demand, the HE string provides energy according to the adjusted current set point, and the additional energy is provided by the HP side.

5.2.3 Brake Condition

Depending on the SoC of the HP string, energy is distributed between the strings. In case of low HP SoC, only the HP side is charged otherwise also the HE side is being charged with a maximum current according to the DC/DC set point.

5.2.4 Depletion Condition

When the HE string drops below the lower SoC limit (e.g. 5 %), only the HP side is used until it is also below the lower limit and the pack is considered completely discharged.

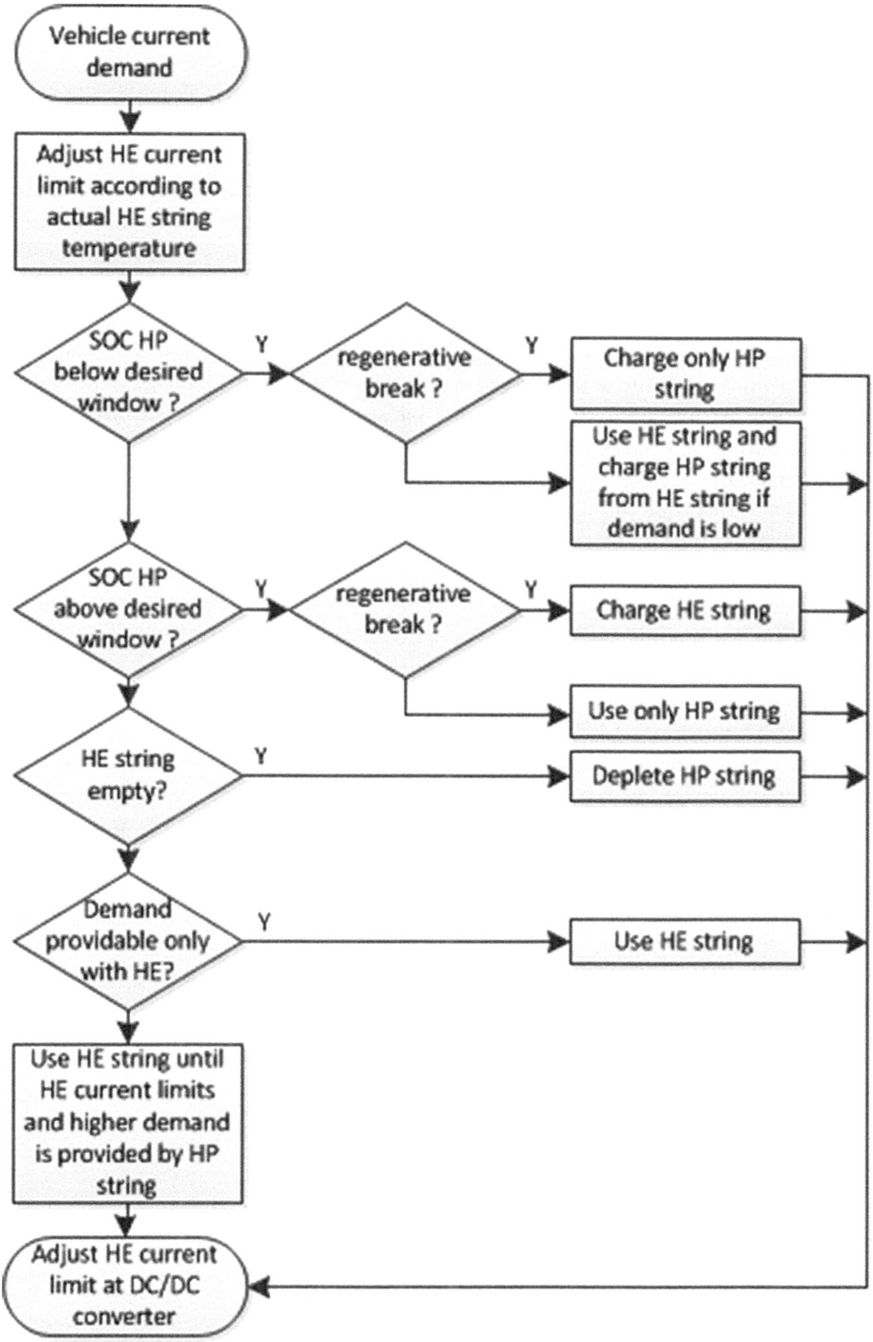

Fig. 11 Energy flow management flowchart

6 Conclusion

The dual-cell battery concept was proposed as an architecture to extend battery life and increase driving range of electric vehicles. The paper describes steps to develop the concept, such as development of the cells, modeling and characterization of the cells, novel temperature sensor development for low cost and highly integrated tracking of all cell temperatures and finally development of the battery management concept, including energy distribution between the high energy and high power cells.

Results are already available from life cycle and energy efficiency testing on the cells developed for the project, and appear to be promising. In addition to test data, outcome of the modeling activities to support advanced battery management system development is presented, along with the latest version of the novel printed temperature sensor. The paper is continued with a description of the final battery management architecture which is a distributed management concept that integrates the DC/DC converter highly in the system electronics. Finally, the approach to energy distribution is described, where the intention is to regulate the power sourced from the energy cells to maximize their life time, and cover the main power requirements from the power optimized cells. Vehicle autonomy can be extended via utilizing a wide SoC range on the energy cells and, making full use of the power cells once the battery pack nears depletion.

The SuperLIB concept will be an enabler for getting higher performance out of today's and tomorrow's Li-ion battery cells via a smart management system, and as such help with increasing the share of electrified drivetrains within the range of options available for our transportation needs. This is an important requirement for keeping our mobility unrestricted for our fossil energy constrained future.

Acknowledgments The authors would like to express their gratitude to the European Commission for financially supporting parts of this research under the 7th Framework Program, Project reference: 285224.

Reference

1. http://www.superlib.eu

SMART-LIC—Smart and Compact Battery Management System Module for Integration into Lithium-Ion Cell for Fully Electric Vehicles

Jochen Langheim, Soufiane Carcaillet, Philippe Cavro, Martin Steinau, Olfa Kanoun, Thomas Günther, Thomas Mager, Alexander Otto and Claudio Lanciotti

Abstract Current limitations of battery systems for fully electric vehicles (FEV) are mainly related to performance, driving range, battery life, re-charging time and price per unit. New cell chemistries are able to mitigate these drawbacks, but are more prone to catastrophic failures due to a thermal runaway. Therefore, new and more advanced management strategies are necessary to safely prevent the energy storage system from ever coming into this critical situation. In this paper, a novel battery management system (BMS) architecture is introduced, which will be able to meet these high requirements by introducing a network that has smart satellite

---

J. Langheim (✉) · S. Carcaillet · P. Cavro
STMicroelectronics, 29 Bd Romain Rolland, 75669 Paris, France
e-mail: Jochen.langheim@st.com

S. Carcaillet
e-mail: soufiane.carcaillet@st.com

P. Cavro
e-mail: philippe.cavro@st.com

M. Steinau
Competence Center Materials and Packaging, Business Unit Transmission, Continental Division Powertrain, Conti Temic microelectronic GmbH, Sieboldstrasse 19, 90411 Nürenberg, Germany
e-mail: martin.steinau@continental-corporation.com

O. Kanoun · T. Günther
Lehrstuhl für Mess- und Sensortechnik Fakultät für Elektro- und, Informationstechnik Technische Universität Chemnitz, TECHNISCHE UNIVERSITÄT CHEMNITZ, Reichenhainer Str. 70, 09126 Chemnitz, Germany
e-mail: kanoun@ieee.org

T. Günther
e-mail: thomas.guenther@etit.tu-chemnitz.de

T. Mager
Department Advanced System Engineering, Fraunhofer Institute for Electronic Nano Systems ENAS, Warburger Straße 100, 33098 Paderborn, Germany
e-mail: thomas.mager@enas-pb.fraunhofer.de

E. Briec and B. Müller (eds.), *Electric Vehicle Batteries: Moving from Research towards Innovation*, Lecture Notes in Mobility,
DOI 10.1007/978-3-319-12706-4_8

systems in each macro-cell or directly in each individual cell. Particular attention will be put on safety and cost issues as well as on 48V application.

Keywords Fully electric vehicle · Battery management system · Lithium-ion battery · Smart cell · Safety · Wireless communication · Electrical impedance spectroscopy · Safety · Cost · Standardization

1 Introduction

SMART-LIC is a battery management project with eight partners funded by the European Commission under the European Green Car Initiative. The project started on May 1st, 2011. Its duration is 42 months.

The partners of this project are STMicroelectronics in charge of both the coordination and the semiconductor side, CRF and Microvett for the automotive side; research representatives from Chemnitz University of Technology (TUC) and Fraunhofer ENAS including the SME Berliner Nanotest especially for battery knowledge; MANZ (former KEMET) for design and production of battery manufacturing equipment and CONTINENTAL for packaging aspects and a strong link in the supply chain between OEM and technology providers.

The objectives of the project are, basically, individual cell management, increasing performance and reducing costs. The latter aspect has largely driven the discussions within the consortium in the first part of the project and has sharpened its view on the system architecture.

One very interesting part is a novel method of battery state determination based on Electro-chemical Impedance Spectroscopy (EIS). Indeed, EIS is an important feature to estimate SoC and SoH for Li-ion batteries, which have a very flat $U_{(SoC)}$ characteristic. Within this project, TUC succeeded to develop an EIS approach that can easily be implemented in a microcontroller.

Furthermore, the project intended to work on communication with the goal of replacing wired communication by wireless communication. Therefore, the consortium has evaluated an adequate frequency range, chosen a sophisticated protocol and investigated different approaches to find the best methods for implementing the antennas into the batteries.

A. Otto
Department Micro Materials Center, Fraunhofer Institute for Electronic Nano Systems ENAS, Technologie-Campus 3, 09126 Chemnitz, Italy
e-mail: alexander.otto@enas.fraunhofer.de

C. Lanciotti
MANZ ITALY SRL, Via San Lorenzo 19, 40037 Sasso Marconi, BO, Italy
e-mail: CLanciotti@manz.com

Packaging of the electronics is a major point for which CONTINENTAL is responsible. One question is for example how to integrate the electronics in a severely challenging environment and where it needs to withstand heating-up of the battery in case of failure. The related reliability aspects are investigated in particular by Fraunhofer ENAS and Nanotest.

Reliability and safety, especially concerning plagiarism, i.e. protecting a potential second user against false information about the remaining health, and thus the commercial value of a battery is also of high importance. Testing and validation is naturally an important task within the project to verify the results of the research work.

Nowadays, battery management systems can be found at systems, modules and partially at the sub-module level, but BMS on individual cell level are still R&D domain. Objective of SMART-LIC was to work especially at a very individual granularity.

2 Highlights of the SMART-LIC Project

2.1 New System Architecture

The main idea was to introduce the distribution of BMS functionalities down to the lowest possible granularity between individual cell and macro-cell level. This new architecture involves advanced concept for active and passive balancing. Wireless communication strategies between satellite and central BMS are considered including EMC issues.

2.2 Improved Battery State Determination

The SMART-LIC battery management concept includes wireless data transfer and electrochemical impedance spectroscopy. The latter has been adapted in order to implement it on automotive embedded solutions with a minimum cost impact. This allows state determination and lifetime prediction on single cell level based on in-system measured data. The implementation of the EIS will also be applicable for improved on-line determination of SoC, SoH and SoF.

2.3 Packaging and Reliability

In general an increase of the temperature of a lithium-Ion cell leads to reductions in lifetime. In the worst case, it can also lead to cell destruction and thermal runaway. SMART-LIC will allow tightening the control of small cell packages and thus

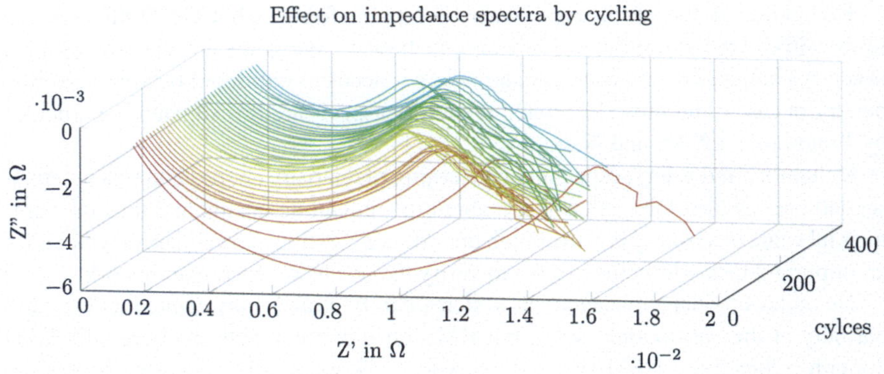

Fig. 1 Highlights of SMART-LIC: electro-impedance spectroscopy characteristics of the battery module

higher exploitation of the battery whilst achieving increased total lifetime. In consequence a reduction of the costs-of-ownership can be expected. The project also aims to create reliable, secure and cost-effective packaging of ECU (BMS module) especially in harsh environments by overmolding.

2.4 48V Application

In order to achieve cost targets in preparation of industrialization and series use, focus was also put on standardization questions. One major input that influenced SMART-LIC just after its start in 2011 was the introduction of the standard of 48V for heavy load on-board power supply and with micro-hybridization a first step towards electric mobility at reasonable costs. Consequently, the project was adapted in order to address this new voltage standard with its topologies allowing use at 48V or multiple of 48V (e.g. 240V) (Fig. 1).

3 Safety Consideration

The evolution of the standardization of safety criteria is one aspect that has influenced the work in SMART-LIC. ISO26262 has been discussed and compiled since about 10 years, but has only come into force in 2011. SMART-LIC has right from the beginning in 2010 considered this new safety standard in order to evaluate its influence on the system design. In the beginning of the project, the required ASIL level for a BMS in electric traction was indicated by different actors in the market to be ASIL B. In the E3CAR project, several new devices were presented compliant with this level. However, during the past years, this requirement has constantly

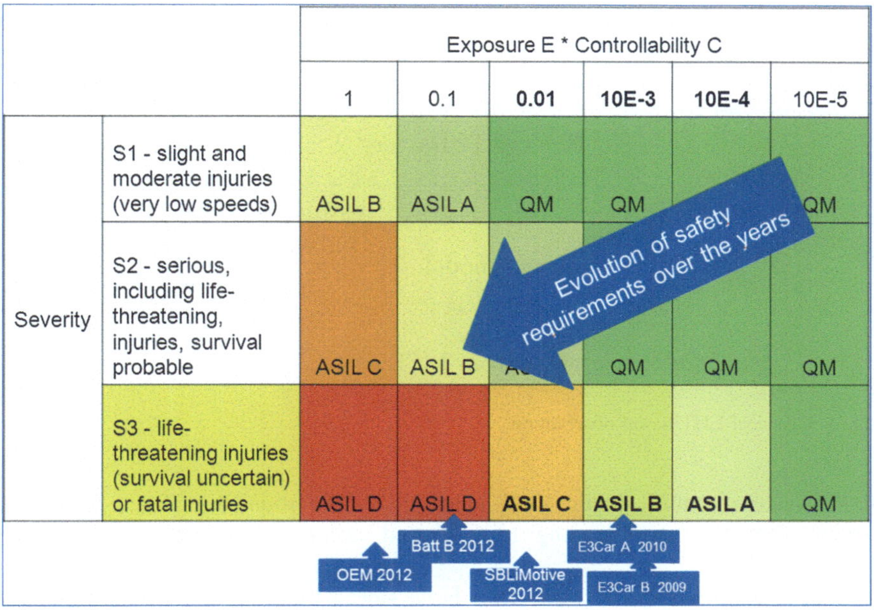

Fig. 2 Exposure and controllability, ASIL determination and evolution of the requirement over time

increased to reach in some cases even ASIL D. Such high level of safety seems exaggerated, but car industry seems to have identified some specific applications that require such a level of safety (Fig. 2).

This had indeed an influence on our view concerning the communication. In the beginning, we considered only one single communication path between the cells that we could have replaced by wireless communication. With a higher ASIL level, redundancy is needed and wireless communication becomes more difficult to realize. In the industry domain we have seen a move to request for double daisy chain communication. This is today's state of the art BMS components.

It is also necessary to do a clear Failure Mode and Effect Analysis (FMEA) in order to regard e.g. the case of over-heated batteries and its consequences for the electronics. It seems that one of the problems of Boeing Dreamliner was heat and electronics not capable to protect the battery due to malfunction.

4 Electro Impedance Spectroscopy (EIS)

For optimal control as well as safe and reliable operation of a battery system, knowledge on different parameters like the state-of-charge (SoC), state-of-health (SoH) or state-of-function (SoF) is crucial. These abstract measures form the base

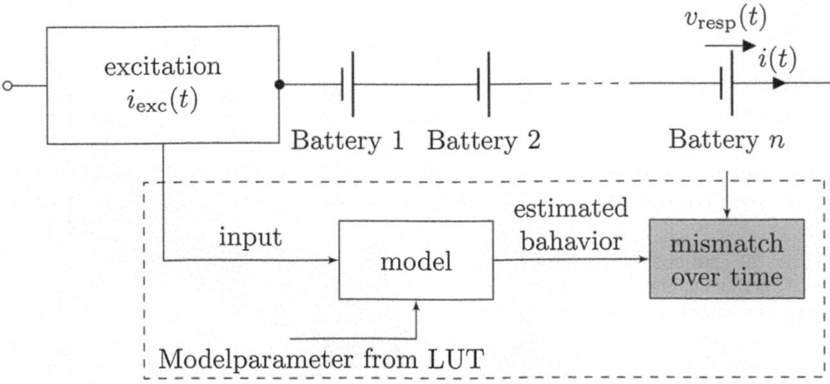

Fig. 3 Classical LUT-based observation

of the system control strategy and soon will be the foundation for a predictive driving schedule, thus leading to increased accuracy requirements on those measures.

Today, battery systems work with large stored look-up tables which contain information on the behavior of a battery over the whole battery lifetime. Due to measurement error accumulation over the lifetime, the quality of information of the state of the battery system based on look-up table values decreases. This consequently makes increased safety margins necessary, which reduces the usable battery capacity and decreases the power limits of the battery (Fig. 3).

To improve this situation, a suitable measurement method is desired. In the laboratory electrochemical impedance spectroscopy (EIS) has proven to be a valuable tool for determining the influence of SoC and SoH on the battery response. EIS presents a non-invasive measurement method to determine the linear system response of an electrochemical system. The electrochemical impedance is obtained by applying a current stimulus on the battery, measurement of the voltage response and subsequent frequency dependent transfer function calculation. The measurement is done over a suitable frequency range allowing for separation of different electrochemical mechanisms inside the battery. To reduce the time required for the measurement and to improve signal to noise ratio a modified multispectral excitation with low crest factor is used (Fig. 4).

Different data analyzing strategies exist with the method of equivalent circuits being the most used one. Fitting parameters of equivalent circuits however can be cumbersome if more sophisticated models are used. The model parameters can be used to determine measures like SoF, SoC or the remaining useful life of the battery. To implement a similar diagnostic method in an embedded system one needs to overcome several obstacles. The excitation must be realizable in an embedded system. Complicated wave-forming hardware is prohibited. The measurement itself must be streamed due to large amounts of data that cannot be stored. An unsupervised data evaluation with high robustness is needed.

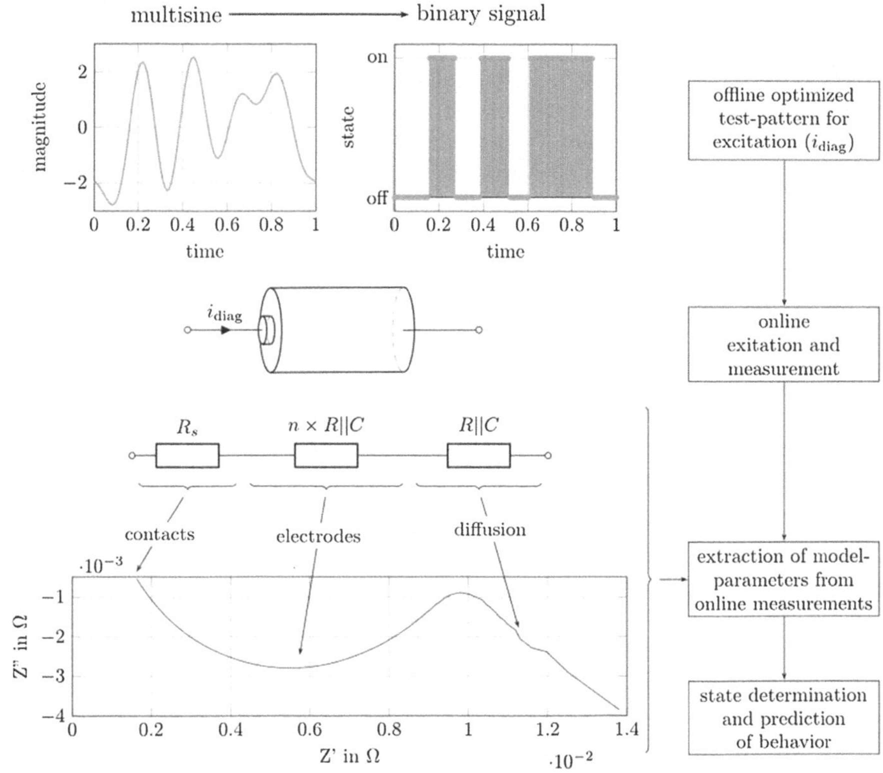

Fig. 4 Determination of system behavior using broadband perturbation signals from measured data

A prototype measurement board was designed for battery diagnosis in an including wave-forming by pulsing technique, analogue frontend for the measurement of very weak, low frequency signals, and the necessary data evaluation.

5 Cost Consideration

The other important aspect was related to costs. Understanding if and how much a more sophisticated electronics can decrease the overall costs in production and operation is a rather complex task. A BMS with active and passive balancing increases the costs. This leads to questions such as how much percentage of the battery costs could be affordable for electronics and what would it bring in cost reduction during the whole battery life. A first cost estimation for our system came up with around 15–20 % of on-costs for a fully developed SMART-LIC system. Discussions with OEMs showed that a system that turned around 15 % costs for the electronics compared to the battery would be realizable. However, in our continuous talks with the industry partners, we met more and more people who requested it should go down to 1 or 2 %. Taking into account the continuous reduction of battery costs over the next years would mean a very tight cost limit for BMS electronics in a traction battery system. This drastically limits tapping the full potential of benefits a BMS could accomplish. Another issue is the knowledge and distribution of knowledge on battery. Until now we could not get a clear indication on the costs generated by longer tests and selection processes in the production or by asymmetric usage in battery packs in cars. Talking to 5 people gave 6 opinions.

Today we depend on the information coming from the battery manufacturers' requirements. However, we do not know where we stand exactly in terms of cost savings.

6 The 48V Standard

Concerning electric and electrified vehicles, there has been a major event right after the start of the project with the promotion of the 48V standard in June 2011. This had also an effect on the direction of discussions in the consortium. This quasi-standard has immediately been taken into account as it might allow increasing

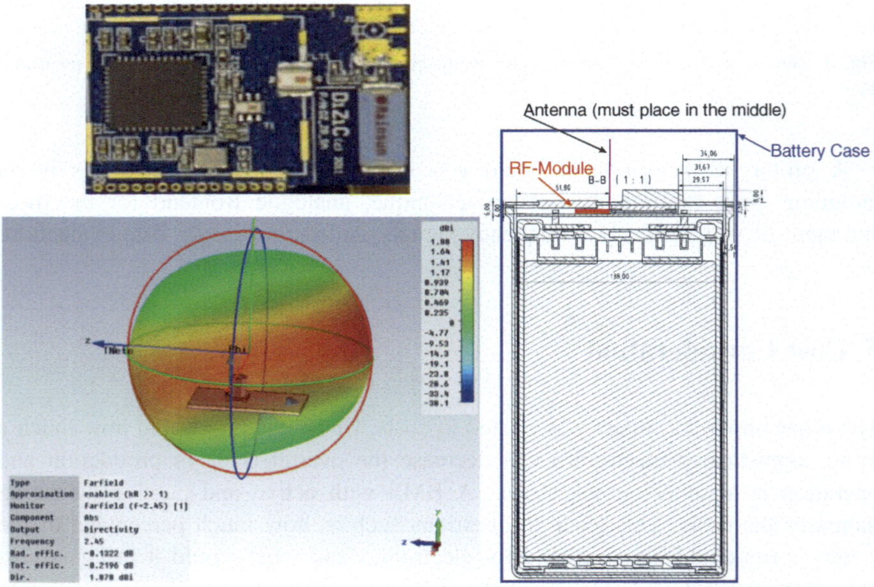

Fig. 5 ZigBee Communication module, position in battery pack, beam characteristic of compact antenna for ZigBee

production volumes and thus reducing costs. This should help increase competitiveness in this new business.

From a 48V-module any further multiple voltages can be derived in analogy (96V, 192V, 384V, etc.).

7 Demonstrators

In conclusion the consortium is focusing on two demonstrators:

- A small module with 4 cells to test R&D activities including wireless communication concepts
- A 48V module to prepare a demonstrator and show the industrial feasibility.

In addition, the wireless communication is standardized. The implemented transceiver chip is 2.4 GHz, IEEE 802.15.4-compliant (Figs. 5 and 6).

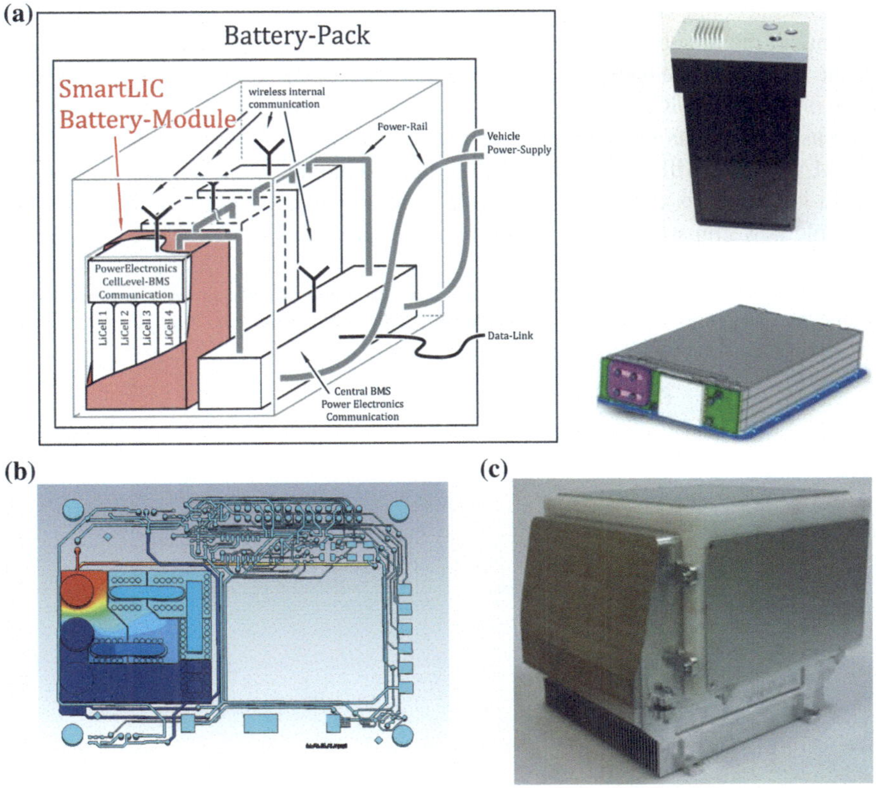

Fig. 6 **a** 1st and 2nd 4-cell demonstrator—SMART-LIC macro cell and example of integration in a battery pack (schematics, CAD design, physical demonstrator). **b** Potential distribution of the power module (4 cell module). **c** First 48V-cell demonstrator

8 Conclusion

The consortium has been struggling, but is progressing and has learnt a lot about batteries in this project.

In conclusion of its considerations, this paper has focused on higher efficiency, higher reliability, higher safety and affordable costs.

In particular, safety considerations (ISO26262) have had a great impact on the design of the electronics and electric concept. This very new standard was not fully understood by the ecosystem of SMART-LIC users. However, during the beginning of the project, the ecosystem became more and more aware of the related requirements (in addition to the incidents in the Boeing 787). This lead to further increased demand concerning functional safety and hence a limitation of choices in the design (wireless communication is less useful in this case).

The influence of cost issues was right from the beginning regarded, but only during the detailed discussions its influence onto the circuit design became clear.

One important aspect in the lifetime management of a battery is the thermal management. This was not part of SMART-LIC, but it became clear that it has to be taken into account in further projects.

Finally, the announcement during the start of the project of the 48V "standard" by the automotive industry has an impact and was detected as such in a very early stage of SMART-LIC. Thus, SMART-LIC is today very happy to be able to present demonstrator results in line with the most recent market requirements.

SMART-LIC will contribute to further understanding of the complexity of BMS for Li-ion batteries and contribute with some interesting new developments, for example new chips inside the ST for battery management.

Acknowledgments The Author would like to acknowledge the European Commission for supporting these activities within the project 'SMART-LIC' (project number: 284879).

The manufacturer's authorised representative in the EU is Springer Nature Customer Service Centre GmbH, Europaplatz 3, 69115 Heidelberg, Germany. If you have any concerns regarding our products, please contact ProductSafety@springernature.com

Printed and bound by CPI Group (UK) Ltd, Croydon, CR0 4YY
01/10/2026
02225989-0002